Topics in Safety, Risk, Reliability and Quality

Volume 38

More information about this series at http://www.springer.com/series/6653

Dan Serbanescu · Anatoli Paul Ulmeanu

Selected Topics in Probabilistic Safety Assessment

Methodology and Practice in Nuclear Power Plants

Springer

Dan Serbanescu
Division of Logic and Models in Science
Romanian Academy
Bucharest, Romania

Anatoli Paul Ulmeanu
Department of Power Generation and Use
Polytechnic University of Bucharest
Bucharest, Romania

ISSN 1566-0443 ISSN 2215-0285 (electronic)
Topics in Safety, Risk, Reliability and Quality
ISBN 978-3-030-40547-2 ISBN 978-3-030-40548-9 (eBook)
https://doi.org/10.1007/978-3-030-40548-9

This Springer imprint is published by the registered company Springer Nature Switzerland AG
The registered company address is: Gewerbestrasse 11, 6330 Cham, Switzerland

To our families

Preface

The PSA studies were initially developed to be used for nuclear power plants, starting from elements of reliability analyses in other areas as for instance aviation. Following the initial period of defining the method, mainly after the TMI accident, PSA methodologies used for NPP became widely spread. PSA is now very well defined by a series of standards. The goal of this book is to present selected topics in PSA, as identified during the last period of more than four decades of use. The book is structure oriented on the PSA tasks, as defined by the standards; it is focused on presenting:

- the Key Topics (**KT**) of the Probabilistic Safety Analysis (PSA) studies. These issues, which arise during the application of PSA standards, are of high interest for PSA practitioners.
- the Problems (**PR**) encountered for the key issues in PSA and
- proposed Solutions (**S**) to the Problems.

The Key Topics are focused on the Main PSA Task, as defined in the standards (Initiating events, event trees, fault trees, etc.).

The Key Topics and the Problems encountered during the implementation of standards and guidance on PSA are focused on the following generic aspects, that are reflected in performing all or most of the tasks in a PSA study:

- limits of applicability, illustrated mainly in problems on processing and using results in each PSA task
- special cases of modelling, as for instance the low frequency events and the plant behaviour under these conditions
- modelling of the combination of various low frequency high impact events in the issue related to the so called 'cliff edge effects'
- interpretation and use of results for risk informed decision making.

The relevance of the Key Topics, which were chosen to be presented in this book, as well as the problems potentially to be encountered in various PSA tasks, is defined by the following criteria:

- the degree to which the issue reflects highly challengeable aspects of modelling NPP as complex systems
- the impact on the use of results for the evaluation of plant safety and risk levels.
- the possibility to use the Solutions in integrated models
- the auditability and stability of possible Solutions to be adopted for the encountered Problems
- the possibility to perform benchmarking of results and to use diverse methods to reach conclusions on the problems.

It is our appreciation that we thank all who have contributed to the preparation of this book.

We also acknowledge the editing and production staff at Springer for their careful and effective work.

Bucharest, Romania
October 2019

Dan Serbanescu
Anatoli Paul Ulmeanu

Contents

Acronyms

BDBA	Beyond Design Basis Accidents
CD	Core Damage
CDF	Core Damage Frequency; Cummulative Distribution Function
CDS	Core Damage States
CEE	Cliff Edge Effects
CET	Containment Event Tree
DBA	Design Basis Accidents
DC	Direct Current source
DEP	Depressurization System
DiD	Defence in Depth
DMP	Decision Making Process
DSA	Deterministic Safety Analyses
DT	Decision Tree
EAL	Emergency Action Levels
EPOWER	Emergency Power batteries
EPS	Emergency Power Supply
EPSA	External Event(s) PSA
ES	End States
ET	Event Tree; Event Trees
FE	Function Event; Function Events
FOAK	First of a Kind
FT	Fault Tree; Fault Trees
HPI	High Pressure Injection
HRA	Human Reliability Analysis
I&C	Instrumentation, Control and alarms
IE	Initiating Event; Initiating Events
IPSA	Internal Event(s) PSA
LER	Large Early Release
LERF	Large Early Release Frequency
LOCA	Loss Of Coolant Accident

LOOP	Loss Of Offsite Power
LPI	Low Pressure Injection
MCS	Minimal Cut Set; Minimal Cut Sets
MUPSA	Multiunit Probabilistic System Assessment
PCS	Primary Coolant System
PDF	Probability Density Function
PIRT	Phenomena Identification and Ranking Table
PSA	Probabilistic Safety Assessment
RBDM	Risk Based Decision Making
RC	Release Category
RHR	Residual Heat Removal
SF	Split Fraction; Split Fractions
SLOCA	Small Loss of Coolant Accident
SMAG	Severe Major Accident Guidelines
SOARCA	State-of-the-Art Reactor Consequence Analyses
SSy	Special Safety Systems
ST	Success Trees
SUA	Sensitivity and Uncertainty Analysis
SUPSA	Single Unit Probabilistic System Analysis

List of Figures

List of Tables

Chapter 1
Introduction

Abstract This chapter is a general introduction to PSA considered from the perspective of the special topics of interest for PSA practitioners and/or of use for the newcomers training in this area. These aspects are mainly related to the following: (a) How the NPP information has to be prepared in order to build a PSA model? (b) Which are the specifics of PSA as a probabilistic method of an NPP analysis versus the deterministic one? (c) Specifics of the PSA method, which are of high impact and importance in complementing the deterministic analyses; (d) A survey of the most important PSA tasks for which there is an interest for practitioners and training of newcomers on how actually to implement various standards provisions. The approach adopted in the book is presented, which consists of describing the main goals and difficulties of the tasks, the proposed solutions (based on the authors' experience) and examples of the use of the suggested solutions.

There are some special features of the NPP as a complex system [1]. In an NPP, the energy from the nuclear fission is transformed into electricity by using, from a thermodynamic point of view, either a two-circuit compound or a one-circuit compound. A schematic representation of a two-circuit NPP is in Fig. 1.1.

The specific features of a mature (well designed and with good operational record) NPP, for which the evaluation of its safety performance may be performed with an acceptable degree of confidence, as defined by standards, are related to some important aspects, as for instance:

- definition of the system boundaries, so that they are well identifiable at any moment in time,
- identification of the important components in various scenarios and of their behaviour,
- definition of the type of interaction between the components and if they comply with the cause–effect law,
- definition of the interdependence matrix between various systems and components during various scenarios,

D. Serbanescu and A. P. Ulmeanu, *Selected Topics in Probabilistic Safety Assessment*, Topics in Safety, Risk, Reliability and Quality 38,
https://doi.org/10.1007/978-3-030-40548-9_1

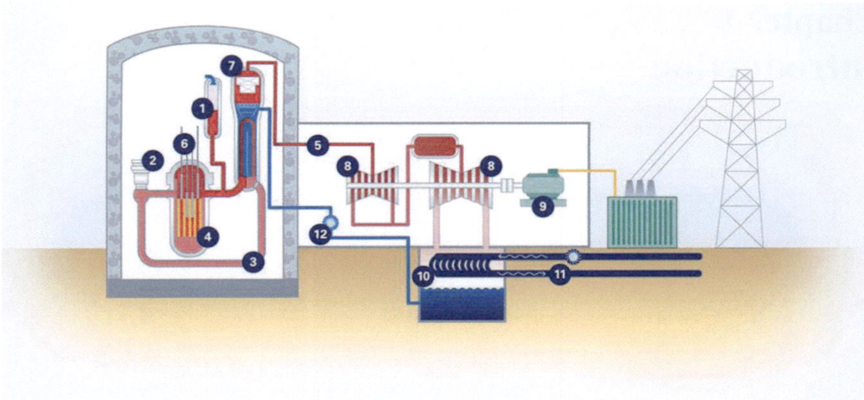

Fig. 1.1 Schematic diagram of two-circuit NPP. **1**—Pressurizer; **2**—reactor coolant pumps; **3**—primary circuit; **4**—reactor; **5**—secondary circuit; **6**—control rods; **7**—steam generator; **8**—steam turbine; **9**—generator; **10**—steam condenser; **11**—cooling water circuit; **12**—feedwater pumps

- evaluation of the degree of correlation between various components behaviour (for one and multiunit cases).

Based on the existing design and operating information for the NPP, a model is built for the purposes of the safety features evaluation. The process of building this model is guided by the answers on the questions to what degree the plant satisfies the highest standards on the issues mentioned above. Throughout this process of model building, the NPP is considered to have the specific features of a complex system (Fig. 1.2).

For a given NPP model, a set of systems have to be defined as a minimum:

- the Special Safety Systems (SSyi), designed for postulated challenges (Initiating Events—IE) and as defined by a set of requirements called Design Basis, which consider plant reaction in case of a series of postulated accidents,
- the Support Systems (Syj) to the special safety systems,
- the Process Systems (SPk), mainly those which may have safety impact and use in some accident scenarios beyond the design basis,
- operational specifics (including Operator Model—OM) and their impact on various scenarios (low power operation, shutdown, etc.) for various lifecycle periods.

An NPP has a high number of systems (on average 200), which are organized into a hierarchical structure and connected between them, so that to fulfill the task of providing energy in a safe mode, i.e. without any adverse impact on the workers, environment and population.

The following features, which are defining an NPP structure are considered important for a PSA methodology:

- hierarchical organization, leading to a hierarchical type of interdependencies, including physical relative positioning,

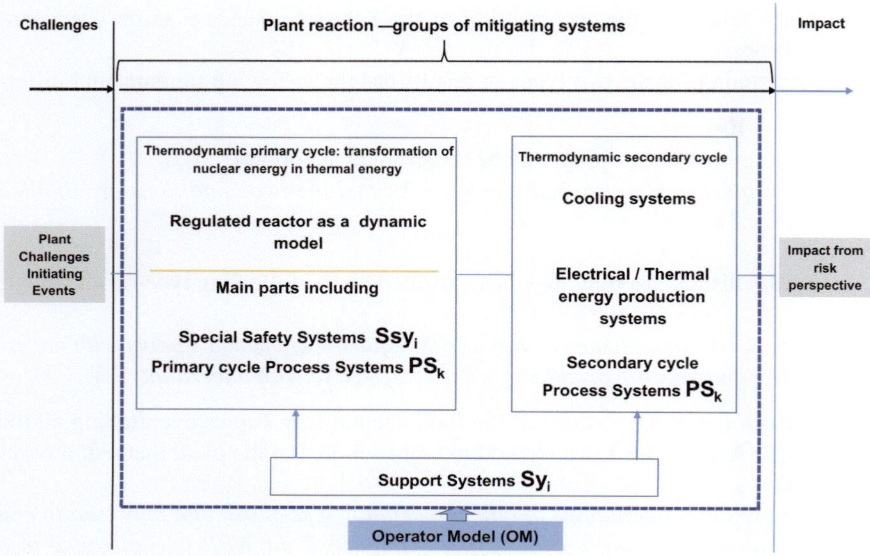

Fig. 1.2 Representation of the NPP reaction to challenges (example: an NPP with two cycles)

- interconnections between systems to assure a specific operational goal,
- combination of various components for a given system in order to fulfill a task defined by design,
- for the identified systems, for each operational state and for each accident case, a set of interdependencies between systems is defined, called System Interdependency Matrix (SIM), as illustrated in Table 1.1.

The NPP impact on the population, environment and workers (Safety Impact) may be evaluated using diverse approaches. They are traditionally divided into two categories:

- deterministic, and
- probabilistic.

However, usually a combination of both approaches and experience gained from operation and/or accidents is used to evaluate. There were also previously presented limitations and areas of applicability for each of the approaches. The main difference between the two approaches resides not in the type of tools used (calculation codes, procedures, etc.), but in the way the results are used as a basis for decisions on

Table 1.1 Sample representation of a Systems Interdependency Matrix (SIM)

	$Ssy1$	$Ssyn$	$SupSyj$	$Spym$
$Ssy1$	–	×	×	
$Ssyn$	×	–	×	
$SupSyj$	×		–	×
$SupSym$		×	×	–

compliance related to the acceptability of the Safety Impact, i.e. in the Decision-Making Process.

The reasoning for the two types of results has the following fundamental difference:

- The deterministic reasoning may be represented as follows:
  If X is requiring Y to produce the effect W and the two conditions are fulfilled then W will take place
  while
- the probabilistic reasoning may be illustrated by the following type of statement:

  Element X known with uncertainty Ux is requiring element Y known with uncertainty Uy and they are producing a known effect W with uncertainty Uw.

The reasoning process is one of the fundamental Key Topics overarching all the tasks in a PSA study. PSA is a method to build a model of the plant that will answer the question :

Which are the combinations of failures, defining a scenario that may lead to end (stable) situations, having a certain Safety Impact, if the NPP is challenged in a certain way?

In order to define the combination of failures, PSA may be used. A set of standards defines PSA and its tasks. In [2–20], it was shown in detail how and why one may consider PSA as a **'triple S' concept** :

- **Structured,**
- **Systemic,**
- **Systematic.**

All those features are the most relevant features of the flow path that defines the PSA. The goal of the PSA is to evaluate the Safety Impact by using a set of criteria called risk metrics.

- **CDF—Core Damage Frequency** defines how the reactor may be damaged, performed in a set of tasks called **PSA level 1**.
- **LERF—Large Early Release Frequency** defines how the containment will fail to release radioactivity to the environment, performed in a set of tasks called **PSA level 2**
- **Risk for population and workers** defines the fatality risks for population (individual and collective) and workers, in a set of tasks called **PSA level 3**.

The risk for an NPP can be described as in formula 1.1.

$$Risk = f(PIE \times PPR \times Pd) \tag{1.1}$$

where
PIE is the probability of the challenge to the NPP, called Initiating Event (IE),
PPR is a probability representing the system pattern for each IE challenge,
Pd is a normalized probability representing the damage produced by a given IE.

For the PSA modelling purposes, the connections are represented in two manners:

- **Event Trees**, as a combination of scenarios describing the successes and failures of some systems designed to cope with the challenges (called Initiating Events—IE). The outcome of each scenario might be either successful to cope with the challenge without an adverse effect or failure to do so. In case of a failure, a set of possible outcomes (defined above as risk metrics CDF, LERF, Risk) takes place.
- **Fault Trees**, as a combination of failures of a mitigating system's components to fulfill its tasks, when challenged in a certain scenario.

By a combination of all mitigating systems failures for the scenarios, leading to an end state of a certain **risk metrics** (**CDF** or **LERF**) , a set of minimal paths to failures (**Minimal Cut Sets— MCS**) is obtained. Summarizing the process described above, the risk metrics are based on a combination of events, which are defining the minimal set of component failures grouped in a set of sequences with the same end state.

The support information to build the PSA model is based on the plant **Model A** (which is describing the energy balances mainly from neutronic and the thermal-hydraulic point of view in a systemic approach). However, the experience of developing PSA so far showed that the use of diverse approaches in modelling the NPP, aside from the operating experience (OPEX) brings very valuable inputs for the risk analyses. Some possible diverse approaches are presented as follows:

- **NPP Model B**—which is describing the NPP by using cybernetic methods,
- **NPP Model C**—which is describing the NPP by considering both energy and entropy losses profiles.

The representation from Fig. 1.3 considers **NPP as a cybernetic machine (Model B)** [21, 22], by using the feedback concepts for the descriptions of the plant, as resulted from the reactor physics and from thermodynamics for such an installation:

- Reactor neutronics (R_1) and the fuel load (RS_1) regulated by the feedback process governed by the delayed neutrons (Fb_1);
- All this part forms the reactor neutronics description for static state, which forms, alongside the thermal hydraulics of the cooling agent and secondary side, the new level of description for the plant, for which the feedback due to the temperature variations impact on the reactor neutronics forms the next feedback chain (Fb_2);
- Finally, the support systems for the neutronics and thermal-hydraulic model of the plant (the dynamic model) are regulated by the next feedback chain (Fb_3).

A cybernetic model of an NPP shows the interconnections and support systems to the reactor as a source of the main risks for the people, workers and environment [23].

Various complementary information about the general design description and cybernetic representation of an NPP may be obtained by considering the thermal-hydraulic model (Fig. 1.4). Figure 1.4 represents an NPP using a Brayton cycle [21, 22]. There is no difference from this thermodynamic modelling point of view between this type of cycle and the more common Rankine cycle. However, the thermodynamic efficiency of a Brayton cycle is much higher.

It is important to mention that, as it was stated even from the main founding PSA methodology documents [2], the risk indications, i.e. high-risk areas in the

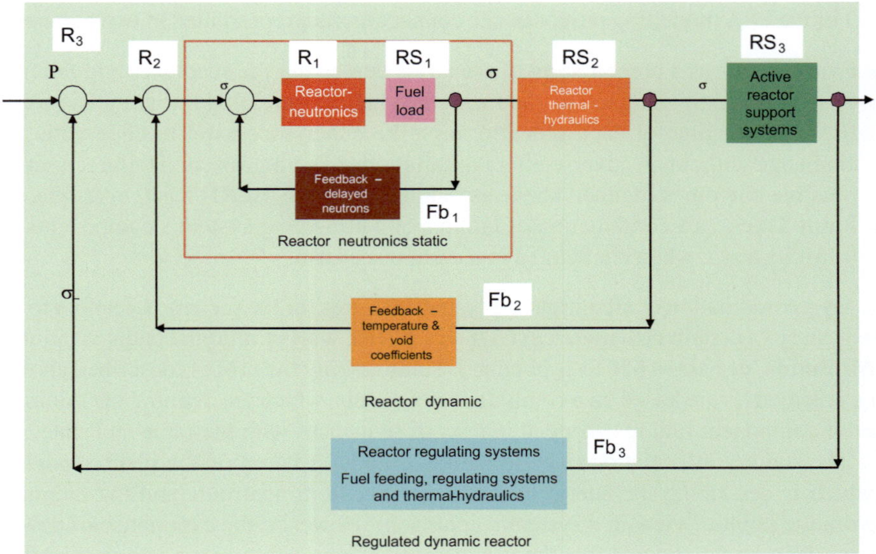

Fig. 1.3 Representation of a nuclear power plant as a cybernetic machine

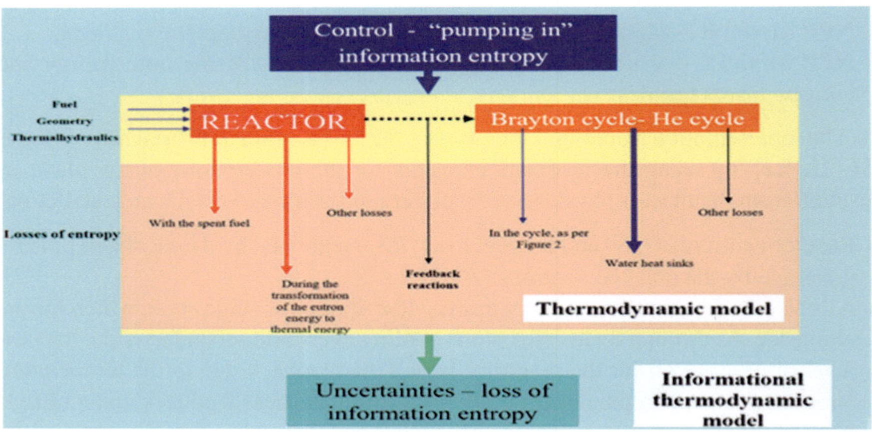

Fig. 1.4 Representation of a nuclear power plant as a thermodynamic machine

plant, may also be obtained using those alternative methods in order to provide inputs to the PSA model. The representations commented so far (general design rule, cybernetic or thermal-hydraulic models of an NPP) provide input to the evaluation of the interdependence matrix of systems (as illustrated in Table 1.1). However, the information from various approaches is complementary and need to be considered as a whole.

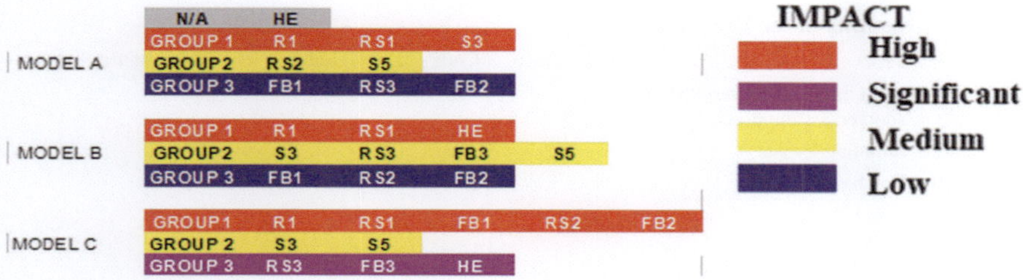

Fig. 1.5 Impact sample for system groups using three models A, B and C

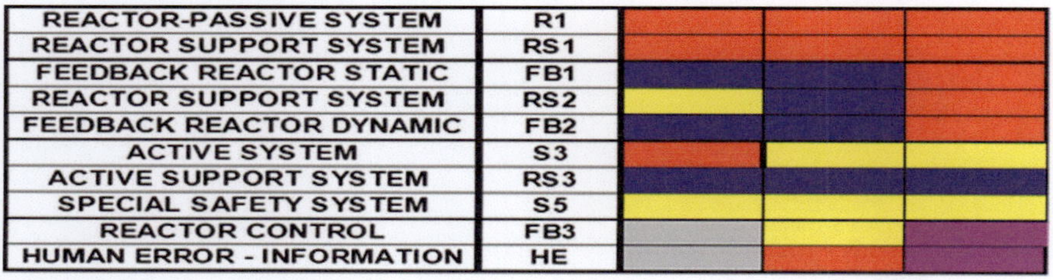

Fig. 1.6 Impact sample for system groups using models A, B and C

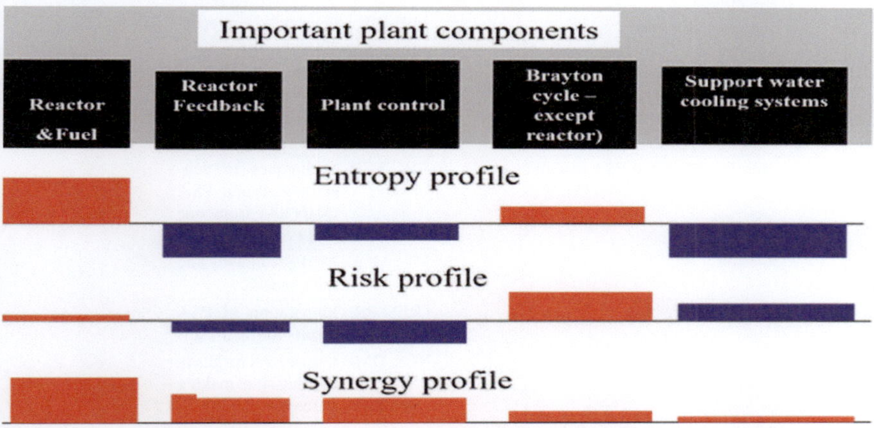

Fig. 1.7 Risk impact evaluation for a nuclear power plant using various methods

The result of the models A, B and C leads to a description of the risk impact of various systems [21, 22, 24]. However, the insights also are related not only to the risk profiles, but also to the profiles of the entropy and synergy (both thermodynamic and information entropies). Figures 1.5 and 1.6 describe the Safety Impact (SI) expected to lead to important risk challenges (notations for the systems as in Fig. 1.3 and its previous description) (Fig. 1.7).

The result of the models provides input on the systemic description of the plant, which is needed to develop **SIM for PSA tasks**. The basic approach used for a *PSA model is to consider a plant as a system of systems, connected between them and*

Table 1.2 Ranking of the PSA tasks

Code	Code	Tasks	Importance	Code	Subtask	Importance
L1	IN	Input information into PSA and adopted assumptions				
	IE	Initiating Events				
	DB	Databases				
	ET	Event Trees				
	FT	Fault Trees				
	IQ	Integration and quantification				
	IQIE	Integration of internal and area or external events in unitary models				
	S&UA	Uncertainty and sensitivity analyses				
L23	GR2	Preparing grouping of the scenarios for the input to PSA level 2				
	ETL2	Specifics of building event trees for PSA level 2				
	L2EP	Use of PSA levels 2 and 3 for Emergency planning technical basis				
LP	DMP	Processing results and preparing for their use in decision making process				
LRE	FB	Feedback to the study		FB_OP	Feedback from operation	
				FB_ST	Feedback from similar studies	
				FB_LES	Implementation of the lessons learnt	
	MU	Building Multiunit PSA models		MU_TCH	Techniques to generate multiunit PSA (MUPSA) from one unit PSA	
				MU_MOD	MUPSA models and their specifics	
	RES	Use of PSA results		RES_PR	Preparation of PSA results for their use	
				RES_DMP	Use of PSA results in the decision making process	
	RSCH	Research topics in PSA methodology				

intercorrelated, so that they will react in a manner to prevent unacceptable Safety Impacts due to any postulated challenge (Initiating Event IE).

In Fig. 1.4, a comparison of three profiles (risk, entropy and energy) is illustrated [21, 22]. As previously shown, the risk metrics obtained using PSA has the advantage of a 'triple S' concept and leads to more detailed and comprehensive results. However, PSA does not exclude the use of various plant models (A, B, C, etc.)

The presentation of the KT, PR and SOL is performed for a set of PSA tasks, listed in Table 1.1.

In Table 1.2, a ranking of the expected impact on performing PSA tasks and subtasks is provided; red indicates a high impact, orange a medium impact and yellow a low impact, but still important for the study.

The following classification and coding is used in the book, which is focused on the aspects guided by the three groups of interest defined before, as follows:

- $Key\ Topics\ (KT)$,
- $Problems\ (PR)$ encountered for a given Key Topic,
- $Solutions\ (S)$ to a problem encountered for a given Key Topic.

For the issues listed before, which are to be presented in the book the following coding system is adopted:

- For the Key Topic = $\mathbf{KT}_x^{TASK}$,
- For a Problem of a KT = $\mathbf{PR}_y^{\mathbf{KT}_x\,x}$,
- For a Solution to a problem of Key Topic = $SOL^{\mathbf{PR}_{yy}\ \mathbf{KT}_x\,x}$.

References

1. Serbanescu D (2015) Selected topics in risk analyses for some energy systems. LAP LAMBERT Academic Publishing
2. PRA Procedures Guide: a guide to the performance of probabilistic risk assessments for nuclear power plants: Chapters 9–13 and appendices A-G (NUREG/CR-2300, vol 2). The American Nuclear Society, LaGrange Park, IL 60525 (1983)
3. NUREG - 1150 : Severe Accident Risks: An Assessment for Five U.S. Nuclear Power Plants. US Nuclear Regulatory Commission, Washington, DC (1990)
4. Defining Initiating Events for Purpose of Probabilistic Safety Assessment. No. 719 in TECDOC Series, International Atomic Energy Agency, Vienna (1993). https://www.iaea.org/publications/981
5. Report NUREG/CR-6172 : Reviewing PSA Based Analyses to Modify Technical Specifications at Nuclear Power Plants. US Nuclear Regulatory Commission, USNRC Washington, DC (1995)
6. Application and Development of Probabilistic Safety Assessment for Nuclear Power Plant Operations. No. 873 in TECDOC Series, International Atomic Energy Agency, Vienna (1996). https://www.iaea.org/publications/5522
7. Regulatory Guide 1.175 : An Approach for Plant specific, Risk-Informed Decision-making: In service Testing. US Nuclear Regulatory Commission, USNRC Washington, DC (1998)
8. Regulatory Guide 1.178 : An Approach For Plant-Specific Risk-informed Decision-making: In service Inspection of Piping. US Nuclear Regulatory Commission, USNRC Washington, DC (1998)
9. Report NUREG/CR-6141 : Handbook of Methods for Risk-Based Analyses of Technical Specifications. US Nuclear Regulatory Commission, USNRC Washington, DC (1998)
10. Living Probabilistic Safety Assessment (LPSA). No. 1106 in TECDOC Series, International Atomic Energy Agency, Vienna (1999). https://www.iaea.org/publications/5820
11. PROCEEDINGS OF THE OECD/NEA WORKSHOP ON SEISMIC RISK, Committee on the Safety of Nuclear Installations PWG3 and PWG5). NEA/CSNI, Nuclear Energy Agency (NEA) / Committee on the Safety of Nuclear Installations (CSNI) (1999). http://www.oecd.org/officialdocuments/publicdisplaydocumentpdf/?cote=NEA/CSNI/R(99)28&docLanguage=En
12. Standard for Probabilistic Risk Assessment for Nuclear Power Plant applications. Nuclear Regulatory Commission / American Society of Mechanical Engineers, ASME, New York (2000)

13. Applications of Probabilistic Safety Assessment (PSA) for Nuclear Power Plants. 1200, International Atomic Energy Agency, Vienna (2001). https://www.iaea.org/publications/6116
14. Specific Safety Guides (2010) Development and Application of Level 1 Probabilistic Safety Assessment for Nuclear Power Plants. SSG-3, International Atomic Energy Agency, Vienna. https://www.iaea.org/publications
15. Attributes of Full Scope Level 1 Probabilistic Safety Assessment (PSA) for Applications in Nuclear Power Plants. No. 1804 in TECDOC Series, International Atomic Energy Agency, Vienna (2016). https://www.iaea.org/publications/10969
16. A guide to Nuclear Regulation in the UK (updated). US Nuclear Regulatory Commission, USNRC Washington, DC (2016)
17. Correlation of Seismic Performance in Similar SSCs (Structures, Systems, and Components). US Nuclear Regulatory Commission, USNRC Washington, DC (2017). https://www.nrc.gov/docs/ML1734/ML17348A155.pdf
18. Guidance on the Treatment of Uncertainties Associated with PRAs in Risk-Informed Decision making, Final Report NUREG-1855. Nuclear Regulatory Commission, U.S.NRC (2017)
19. PSA ASAME (2017) Methodology for Selecting Initiating Events and Hazards for Consideration in an Extended PSA, Nuclear Fission: Safety of Existing Nuclear Installations, Work-Package WP30/D30.7/2017-31. EU: Seventh Framework Programme
20. United States Nuclear Regulatory Commission (1975) Reactor safety study. An assessment of accident risks in US commercial nuclear power plants. http://inis.iaea.org/search/search.aspx?orig_q=RN:35053391
21. Serbanescu D (2003a) Risk, entropy, synergy and uncertainty in the calculations of gas cooled reactors of PBMR type. https://www2.scopus.com/inward/record.uri?eid=2-s2.0-84933178247&partnerID=40&md5=b9fd8f10427aa074f780b50d6139975b
22. Serbanescu D (2005) Some insights on issues related to specifics of the use of probability, risk, uncertainty and logic in PRA studies. Int J CritAl Infrastructs 1(2–3):281–286. https://doi.org/10.1504/IJCIS.2005.006124
23. Health & Safety Executive (2001) Reducing Risks, Protecting People. www.hse.gov.uk/risk/theory/r2p2.pdf
24. Some specifics of the use of probabilistic risk analyses as a support to the evaluation of safety margins and the interface with the deterministic based decisions. In: Proceedings of the Technical Meeting on Effective combination of deterministic and probabilistic safety analysis in plant safety management, Paper 29, IAEA (2006). https://doi.org/10.13140/RG.2.1.2794.8647

Chapter 2
Special Topics in Probabilistic Safety Assessments (PSA) Level 1

Abstract The special topics presented in this chapter are related to the in Probabilistic Safety Assessments (PSA) level 1, which is evaluating the risk impact considering that the reactor is damaged. The tasks are presented in the approach mentioned in the introduction and in the order of their flow path during the performance of such a study, i.e.: (a) How the input to the PSA model is prepared and which are the main challenges? (b) The screening of the hazards and the evaluation of the considered challenges to the NPP (Initiating Events), possibly leading to risk increase; (c) The development of the databases for failures of components and frequencies of the initiators; (d) Description of the plant reaction to the challenges by modeling it in a series of event trees for the chosen list of initiators in the previous tasks; (e) Description of plant barriers to the challenges identified in the event trees; (f) Integration and quantification general approach and special aspects of this task for internal/area or external events in NPP unitary models. Uncertainty and sensitivity analyses of PSA level 1, as basic information for further use of results in the decision-making process.

The Key Topics in a PSA level 1 ($\mathbf{KT}_L^1$) are related to the following:

- Input information into PSA and adopted assumptions (IN),
- Initiating Events (IE),
- Databases (DB),
- Event Trees (ET),
- Fault Trees (FT),
- Integration and quantification (IQ),
- Integration of internal and area or external events in unitary models (IQIE),
- Uncertainty and sensitivity analyses (USA).

The best recommended practice for building PSA models in order to optimize their size and perform easy reviews and corrections further on is to develop it in a step by step, structured, hierarchical approach. The Plant model used for the evaluation

D. Serbanescu and A. P. Ulmeanu, *Selected Topics in Probabilistic Safety Assessment*, Topics in Safety, Risk, Reliability and Quality 38,
https://doi.org/10.1007/978-3-030-40548-9_2

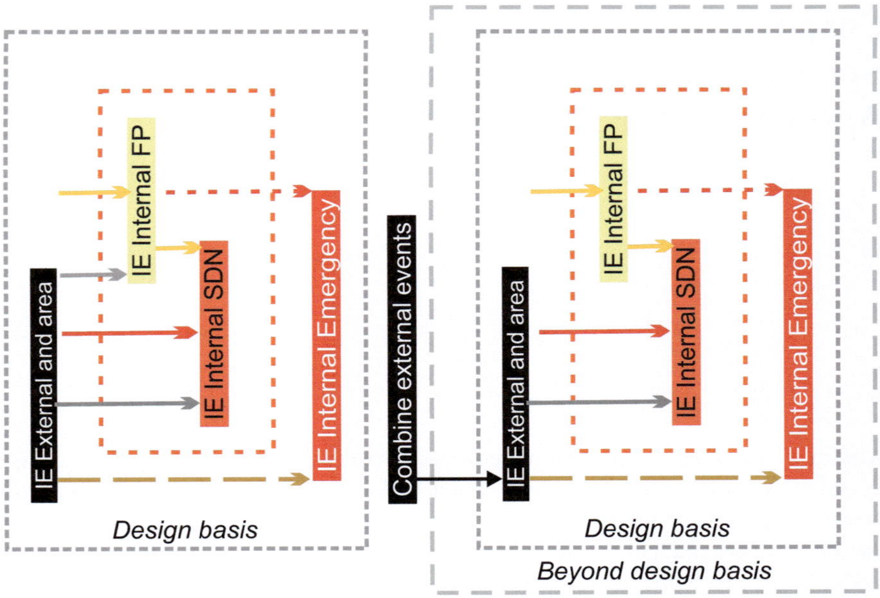

Fig. 2.1 Reactor model in layers for DBA and BDBA

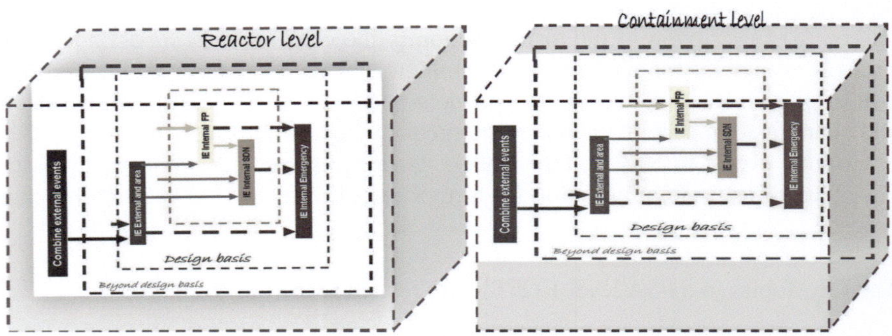

Fig. 2.2 Reactor and containment levels in successive layers for DBA and BDBA

by using the PSA methodologies consists mainly of the following features (Figs. 2.1 and 2.2) [1]:

- The model is developed in the first step for the reactor itself and then for the containment.
- The development for each of the NPP parts is done in layer upon layer of models built on one another using special techniques, in order to optimize its size.
- For each part of the NPP, the steps are as follows:

– First, the model for the Design Basis Accidents (DBA), for which a set of postulated DBA IE list is built. The model starts with a list of challenges (IE) due to internal failures (Internal IE). Based on this initial model the new challenges due to area events (Area IE—fire, flood from internal sources) are included, using a set of logical connectors. To the model is then added the part describing NPP's reaction, caused by the external challenges (External IE) leading to a level of plant reaction within the envelope of the DBA. DBA envelope is mainly defined by deterministic analyses and confirmed by operating experience (OPEX);
– Starting from the model of DBA, the NPP's reaction at the events of severe impact type, beyond the DBA (BDBA), is added. BDBA IE are mainly external events of catastrophic nature. These events are low-frequency high-impact events, related to the so-called 'Cliff Edge Effects' (CEE);
– The process is developed in a step-by-step manner and the layers are added by using previous layer and adding logical connectors. The logical connectors (called Switches) are adding plant reaction modules for the new layers and making corrections to the previous layers so that they will correspond to the new set of challenges (from internal to area and then to external IE).

The total NPP model for the reactor and the containment parts, considering the layers of various types (for the Internal IE, for the Area IE and for the external events for DBA and for the BDBA may be evaluated using a series of combinations for the calculation of the risk metrics. The layers of the model for a given part (reactor or containment) are marked in Fig. 2.3 by '1'–'4'. The possible combinations of the

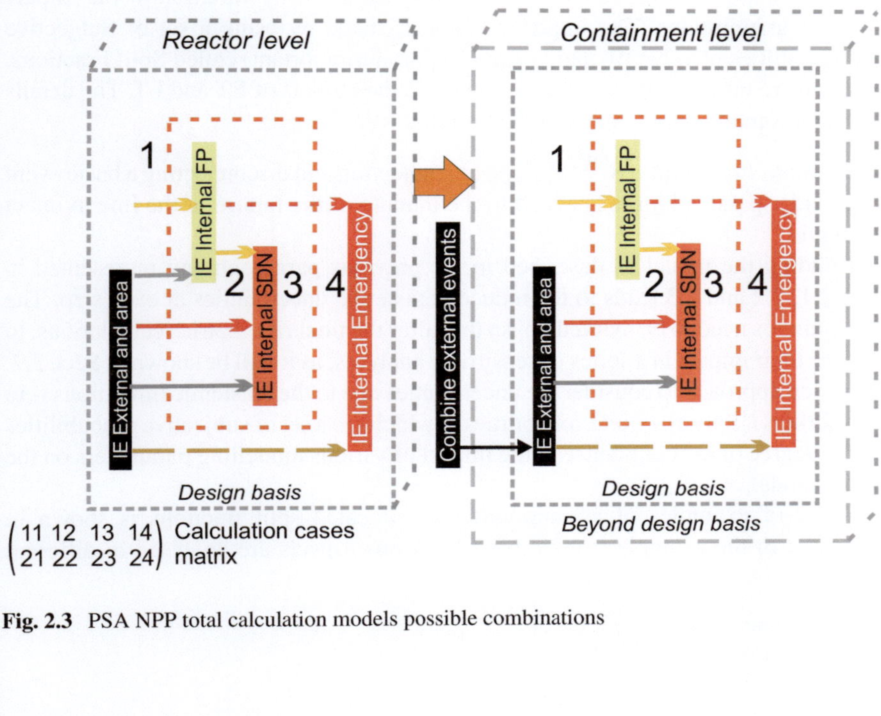

Fig. 2.3 PSA NPP total calculation models possible combinations

resultant model, depending on the PSA objectives in a given study, are also indicated in the matrix shown in Fig. 2.3.

The PSA model itself may be also considered as a space of states defined in an algebra structure for the tasks mentioned above. The impact of this approach is shown in Sect. 2.6.

2.1 Input Information into PSA and Adopted Assumptions

The input to the PSA starts with the knowledge of the plant design and operating documentation. For the unknown or uncertain aspects, a set of assumptions are defined.

- *The Key Topic for input information* ($\mathbf{KT}_1^{IN}$) is to define and consider in the results the impact of the initial input uncertainty of epistemic nature.
  Example: Not existing plant-specific databases for considering the failure of passive components in a SSy (supports, piping systems, etc.).
- *Problem for the* $\mathbf{KT}_1^{IN}$ ($\mathbf{PR}_1$ $\mathbf{KT}_1$) is how to quantify, review and consider the impact of the initial input of the epistemic uncertainties in the final PSA risk metrics results.
  Example: Databases with limited information on passive components (for instance, supports or piping failures in one SSy—Special Safety Systems).
- *Solution for the* $\mathbf{PR}_1$ $\mathbf{KT}_1$ ($\mathbf{S}_1$ $\mathbf{PR}_1$) is to assume from the beginning that, there will be a series of models to be developed for PSA, by variation of the impact of the assumptions. The implementation is made by using a set of Subjective Probabilities of value '0' and '1' (not important/important) called Split Fractions, which are introduced from the beginning in the models of ET and FT. The details of this solution are included in the Solution from ET.

Example 1 of solution $\mathbf{S}_1^{\mathbf{PR}_1}$: A logic of connecting and disconnecting a basic event to consider passive failure is added to the module of the failure of the line to inject the liquid.

Building the model as described in the previous paragraph and represented in Figs. 2.1, 2.2 and 2.3 leads to the induced epistemic uncertainties at each step. The uncertainties need to be marked up, so that after the final risk metrics calculations, to achieve their impact in a series of sensitivity analyses, as it will be shown in Sect. 2.7. One good approach to consider the uncertainties due to the epistemic limitations is to use markers ('Split fraction Probabilities'), which are a set of subjective probabilities on the degree of the confidence in the impact of various modelling parameters on the whole model.

A set of examples of assumptions and allocated split fractions is shown in Table 2.1; in the example from Table 2.1 various aspects are analysed in different types of SUA:

- *SUA of one variable—SUA 1* or
- *SUA of more variables—SUA 2*,

Table 2.1 Example of split fractions prepared for sensitivity cases [1]

Code	Descriptions Assumptions	SUA1	SUA2
IE_SF_CONT_CASES	IE SF containment cases		
SF_IMPACT0_IE_CONTS0	Split fraction impact very low Containment (CONTS) state type 0 (leak-tightness very low affected)		
SF_IMPACTE_IE_EMERG	Split fraction impact due to emergency cases		
SF_IMPACT0_IE_SLOCAN	Split fraction impact due to Small Loss of Coolant Accident (SLOCA/Transient) with Primary Coolant System (PCS) available		
SF_IMPACT1_IE_CONTS1	Split fraction impact very low CONTS state type 1 (Leak-tightness Low affected)		
SF_IMPACT1_IE_SLOCAN	Split fraction impact 1 due to SLOCA/Transient w/o PCS in design basis		
SF_IMPACT2_IE_CONTS2	Split fraction impact very low CONTS state type 2 (leak-tightness Medium affected)		
SF_IMPACT2_IE_MLOCAN	Split fraction impact 2 due to MLOCA in design basis		
SF_IMPACT3_IE_CONTS3	Split fraction impact very low CONTS state type 3 (leak-tightness High affected)		
SF_IMPACT3_IE_LLOCAN	Split fraction impact 3 due to LLOCA in design basis		
SF_IMPACTE_EMERGENCY	Split fraction impact due to emergency situation		

where the Split Fraction impact might be related to Table 2.1 and defined for the steps in Figs. 2.1, 2.2, 2.3, as, for instance:

- IE for reactor and containment (cont) cases,
- IE induced during emergency cases on another unit on site,
- Very low containment (CONTS) state of low level of impairment (type 0),
- Small Loss of Coolant Accident (SLOCA)/Transient with Primary Coolant System (PCS),
- CONTS state of medium level of impairment (type 1),
- CONTS state for high level of impairment (type 2),
- Medium Loss of coolant accident (MLOCA) In DBA
- CONTS state for catastrophic (BDBA) level of impairment (type 3)
- Large loss of coolant accident (LLOCA) in DBA
- Emergency situation.

Table 2.2 Example of split fractions prepared for sensitivity cases [1]

ID	IE seismic					Description
	IE_SE_1	IE_SE_2	IE_SE_3	IE_SE_4	IE_SE_5	
FE_SF_IMPACT0	X					Function Event Split fraction impact 0
FE_SF_IMPACT1		X				Function event split fraction impact 1
FE_SF_IMPACT2			X			Function event split fraction impact 2
FE_SF_IMPACT3				X		Function event split fraction impact 3
SF_PS_DC_DEP				X		Function Event Split Fraction Power Switch no DC for DEP(depressurization system)
SF_PS_DC_HPI				X		Function Event Split Fraction Power Switch no DC for HPI (High Pressure Injection System)
SF_PS_DC_LPI				X		Function Event Split Fraction Power Switch no DC for LPI (Low Pressure Injection System)
SF_S_IA_DEP				X		Function Event Split Fraction Switch no IA for DEP
SF_S_IA_HPI				X		Function Event Split Fraction Switch no IA for HPI
SF_S_IA_LPI				X		Function Event Split Fraction Switch no IA for LPI
SF_S_SW_DEP				X		Function Event Split Fraction Switch no SW for DEP
SF_S_SW_LPI				X		Function Event Split Fraction Switch no SW for LPI

Example 2 of solution **S**_1_**PR**_1: If the initiator is of special type (as, for instance, the seismic initiator E_SE_x), then a special technique is used to consider it as acting in levels (In Table 2.2 a set of 5 levels are represented). However, in this case, one might expect that a series of epistemic uncertainties in the evaluation of the IE_SE_x on the NPP has to be considered. Table 2.3 represents the possible epistemic markers for this case.

SF for external events in the function events of scenarios:

- impact 0—very low,
- impact 1—low impact,
- impact 2—medium,
- impact 3—high.

Split fraction for external events (seismic) to switch modules in the model:

Table 2.3 Example of *Tsunami IE Impact Matrix (IM) on NPP—Internal IE triggered by Tsunami IE*

Tsunami IE group	Tsunami heigth exceedence (m)	Internal IE affected by Tsunami IE					
		IE_LODC_FP	IE_LOAC_FP	IE_LOSP_SDE	IE_LOSW_FP	IE_SLOCA_FP	IE_TRAN_
T1	3-5		X	X			X
T2	5-7	X	X	X	X	X	X
T3	7-9	X	X	X	X	X	X
T4	9-12	X	X	X	X		X
T5	12-15						X

- Power Switch no DC (Direct Current source) for DEP (Depressurization System),
- Power Switch no DC for HPI (High-Pressure Injection),
- Power Switch no DC for LPI (Low-Pressure Injection),
- Switch no IA for DEP,
- Switch no IA for HPI,
- Switch no IA for LPI,
- Switch no SW for DEP,
- Switch no SW for LPI.

2.2 Initiating Events

The input to the PSA starts with the definition of the list of challenges to the NPP (list of initiating events IE).

- the Key Topic for the Initiating Events (**KT**_2 IE) is to have a list of IE that is representative and complete for the PSA model of the given NPP.
  Example: Given a list of IE for a new PSA type decide if it is representative and complete.
- Problem for the **KT**_2 IE (**PR**_2 **KT**_2) is how to evaluate if a list of IE is complete.
  Example: Evaluate the completeness of the list in Fig. 2.4.
- Solution for the **PR**_2 **KT**_2 (**S**_2 **PR**_2) is to use a procedure for the completeness review of an IE list.

Example 1 for **S**_2 **PR**_2: A procedure for IE [1] completeness review based on a failure mode evaluation. One possible approach to review the IE completeness is based on a failure mode evaluation of important systems. For this purpose, a graph representing the failure scenarios for the system is built (this scenario might be in a Fault Tree (FT) format).

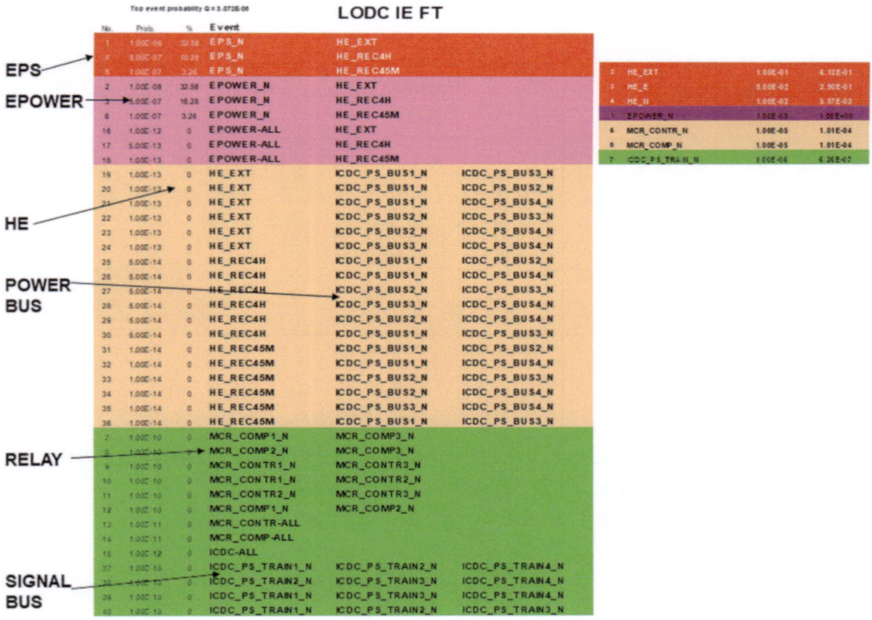

Fig. 2.4 Procedure IE

The sample FT developed for the failure of the DC (battery systems) in an NPP leads (for a particular case) to a series of dominant failure scenarios (MCS) illustrated in Fig. 2.4. The failure scenarios might be grouped in the following dominant ones for systems:

- Emergency diesels (Emergency Power Supply—EPS) and Emergency (batteries) power (Epower)
- Support cooling systems to the inverters to the batteries (Power buses, Relays of the electrical part and Instrumentation, control and alarms—I&C systems).

From the IE FT in Fig. 2.4, the conclusion is that failures of the emergency power systems and failures of the cooling and I&C systems have to be included in the list of IE for the particular case under review. In this case, the calculation of the input data to the IE is performed considering the generic approach adopted as part of the Database task in PSA.

However, there are several important issues to be considered:

- *The failure probabilities* will be considered similar to all the other distribution for the whole PSA, which is usually *a log-normal distribution*
- *The calculations are performed by using the medium values. However special SUA techniques are available in PSA to consider uncertainties*, as shortly mentioned in the Example 1 of solution $\mathbf{S}_1^{\mathbf{PR}_1}$ and which is detailed in Sect. 2.1.

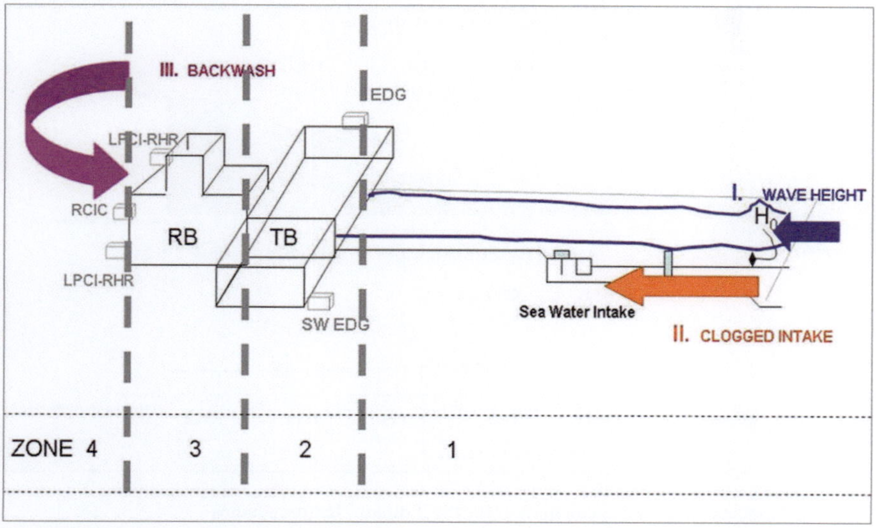

Fig. 2.5 Schematic description of the tsunami impact on an NPP

Example 2 solution **S**_2 **PR**_2: A review of the completeness of IE for external events considering external events initiator, as, for instance, tsunami. Tsunami generates three types of IE due to the following effects on the NPP (Fig. 2.5) [1]:

1. Wave height exceedance over the postulated maximum height for the site.
2. Sand clogging of the water intakes to the NPP.
3. Backwash of the debris from the plant to the sea after the first hydraulic impact ended. and the site was flooded.

From all the tsunami groups of IE, only the first one might be described by using the probabilistic techniques of PSA type, the other two are described by deterministic analyses, which do not differ as a technique against types of IE described in *Example 1 of solution* $\mathbf{S}_1^{\mathbf{PR}_1}$.

For the description of such IE, special techniques were developed initially they were developed for seismic initiators and presently work is performed to use the same techniques for the *Tsunami of type 1 IE*.

The main aspect of the evaluation of Tsunami IE type 1 is due to the fact that the effect on the plant has to be evaluated considering that a failure is a result of a combination of two probabilistic type events (Fig. 2.6):

- An event described by *the probability that the wave height will exceed the reference value defined as a limit for a given site, called hazard (H(h))*, and
- An event described by the *probability that certain elements/systems of the NPP will fail if 'the probability of tsunami wave' will exceed the safety limit. This probability is called Fragility (F(h))*, describing the manner the NPP systems and components may deteriorate during the first event.

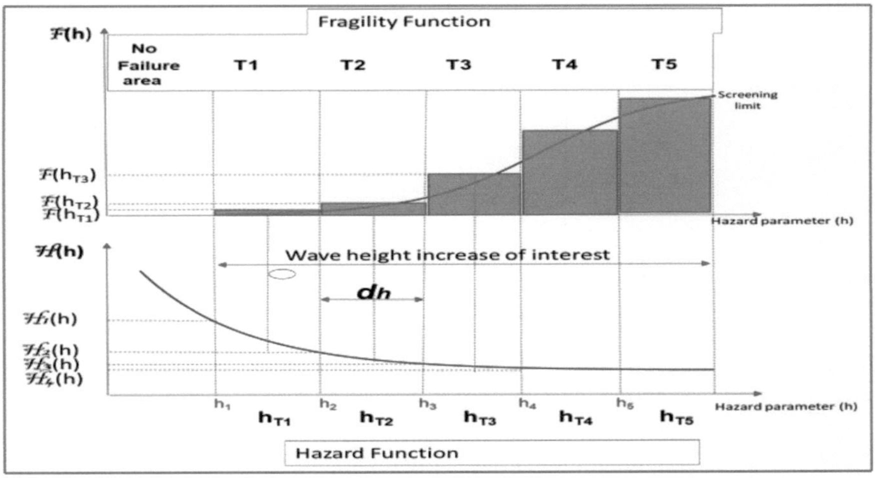

Fig. 2.6 Representation of the calculation for the Tsunami IE frequencies

Both $F(h)$ and $H(h)$ are functions on the magnitude of exceedance of the wave height over the defined limit for the site (h). The combination of the two probabilities consists of calculating the convolution integral. A simplified calculation of the convolution integral, based on the existing approach is described by the formula (2.1) and presented in Fig. 2.6. In formula (2.1), the probability of failures due a tsunami event resulting from the approximate calculation of the convolution integral is P_F:

$$P_F = \int_{h_1}^{h_5} H'(h)F(h)dh \tag{2.1}$$

Summarizing the IE of tsunami type will be split into a series of IE (IE T1 to IE T5 (Fig. 2.6). Each tsunami IE will generate a series of internal IE (Table 2.3) depending on the fragilities of various systems of NPP, as, for instance:

- Plant transients (ATWS) at full power
- Large Loss of Coolant Accidents (LLOCA) at full power
- Loss of Offsite Power (LOOP) at full power, in shutdown, etc.

Further evaluation of the impact of tsunami IE on the NPP is performed by considering the Impact Matrix from Table 2.3.

As it is detailed in the paragraph related to the specific issues for the Event Trees (ET), an important aspect is to define the barriers assumed to protect the NPP in a given scenario for each ET developed for the Internal IE.

These barriers are called hereafter *Function Events (FE). FE are actually FT that describe barrier reaction to cope with a certain challenge.*

In the case of Tsunami IE triggering internal NPP IE, the FE have some parts deactivated and other new parts (Table 2.4).

Table 2.4 Sample of an IE Tsunami Interdependence Matrix (IM) with the Function Events (FE) [1]

		IE tsunami height / run-up						
No.	ID	IE_T1	IE_T2	IE_T3	IE_T4	IE_T5	IE_T6	Description
1	FEA_CREC_SDE				x			Function Event AC recovery at emergency shutdown
2	FEA_LTHEAT_SDE			x				Function Event alternate heat sink in emergency shutdown
3	FE_CDS_SDE			x				Function Event condensate injection in emergency shutdown
4	FE_CHR_N		x					Function Event Containment Heat Removal in emergency shutdown
5	FE_CRED_FP		x					Function Event Control Rod Drive at Full Power
6	FE_CV_N		x					Function Event Containment Ventilation
7	FE_DEP_FP		x					Function Event Depressurization at Full Power
8	FE_DEP_SDE			x				Function Event Depressurization in emergency shutdown
9	FE_DETCP_LOCA_FP		x					Function Event detection of LOCA outside primary containment at Full Power
10	FE_DIAG_SDE			x				Function Event no diagnose loss of shutdown cooling in emergency shutdown
11	FE_EAC_FP		x					Function Event Emergency Power AC failure at Full Power
12	FE_EC_FP		x					Function Event Early Containment Control at Full Power
13	FE_EPS_SDE				x	x		Function Event Emergency Power available at emergency shutdown
14	FE_HPCI_SDE				x	x		Function Event to flow from CRD in emergency shutdown
15	FE_HPI_FP			x				Function Event High Pressure injection at Full Power
16	FE_INH_FP			x				Function Event Inhibition of Automatic Depressurization System and LVI control failure at Full Power
17	FE_INTCP_N			x				Function Event Primary Containment Integrity
18	FE_ISO_LEAK_SDN		x					Function Event Isolation of leak at normal shutdown
19	FE_LI_N		x					Function Event late injection
20	FE_LPI_FP		x					Function Event Low Pressure injection at Full Power
21	FE_LPI_SDE				x	x		Function Event Low Pressure injection at emergency shutdown
22	FE_OVER_SP		x					Function Event Overpressure protection at Full Power

Table 2.5 Sample results for Tsunami IE frequencies and the frequencies of the Internal IE induced by them [1]

Tsunami IE group	Frequency	Probability of Failure					
		IE_LODC_FP	IE_LOAC_FP	IE_LOSP_SDE	IE_LOSW_FP	IE_SLOCA_FP	IE_TRAN_
T1	0.001214	0	1.02E-04	7.26E-04	0	0	7.26E-04
T2	0.008649	1.01E-04	2.66E-04	6.92E-04	1.22E-03	6.38E-06	6.92E-04
T3	1.14E-04	2.98E-05	5.05E-05	7.86E-05	7.53E-05	5.26E-06	7.86E-05
T4	1.69E-05	9.18E-06	1.23E-05	1.51E-05	1.31E-05	0	1.51E-05
T5	4.22E-06	0	0	0	0	0	4.07E-06

The calculation of the input data to the IE for external events of tsunami type is performed as follows:

- The frequencies of Tsunami IE and of the internal IE are calculated for each group considering the evaluation as defined by the convolution integral (formula (2.1) and Fig. 2.6); a sample case is presented in Table 2.5.
- The calculations for the part of the internal PSA model, which are included in the external event model, are performed as per the standard database methodologies (For some specific features in this case, see Example 3 solution **S**_2 **PR**_2).

A representation of the process described before is in Fig. 2.7, with an illustration of a Tsunami ET connection with Internal IE LOOP in Figs. 2.8 and 2.9 [1].

Example 3 solution **S**_2 **PR**_2: Sample list of internal IE (as referenced in Tables 2.3 and 2.4). IE are grouped, mainly by considering the safety functions triggered by them (Table 2.6). The calculation of the frequencies is following the general database rules for the PSA tasks. IE have different impact and frequencies if the NPP is in various operating states (Full Power—FP or Shutdown—SDN, for instance). On the other side, the data for IE are checked against the history of the NPP and/or similar plants (as part of the OPEX tasks) and also considering the information from safety and operating documents.

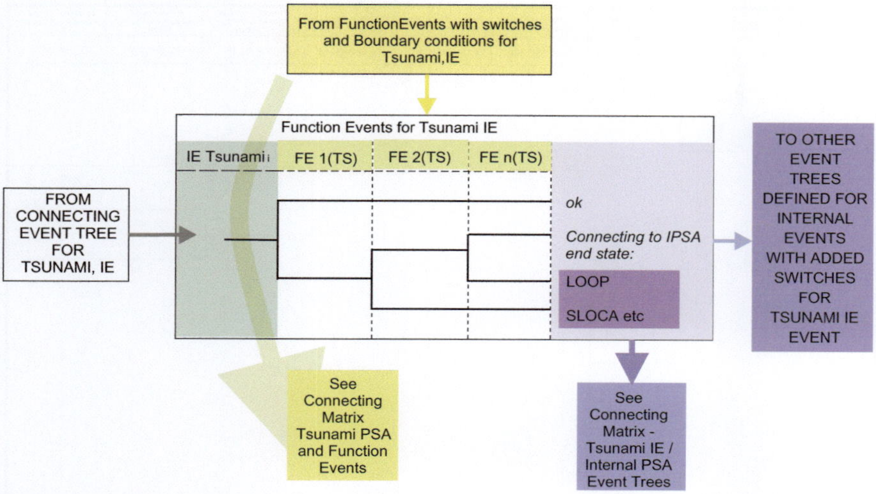

Fig. 2.7 Sample representation of the connection of the model for a Tsunami IE with the internal IE PSA model

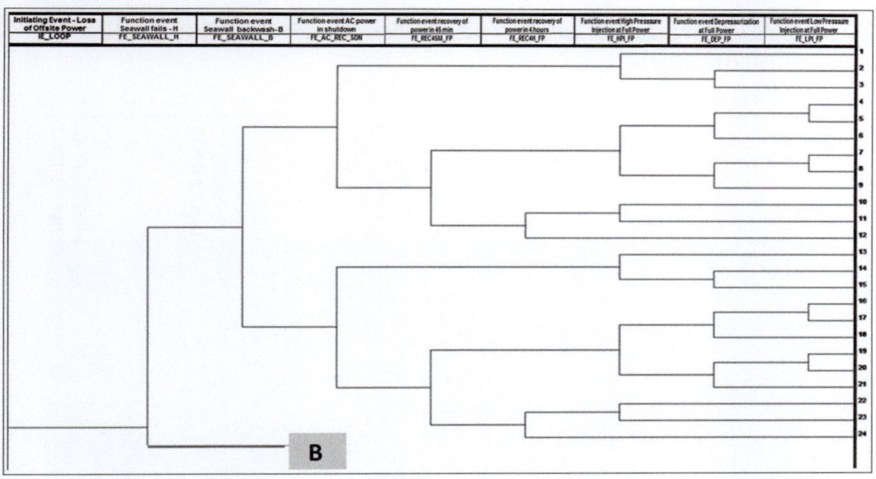

Fig. 2.8 Sample of an ET for Loss of Offsite Power (LOOP) connected to the Tsunami IE (1)

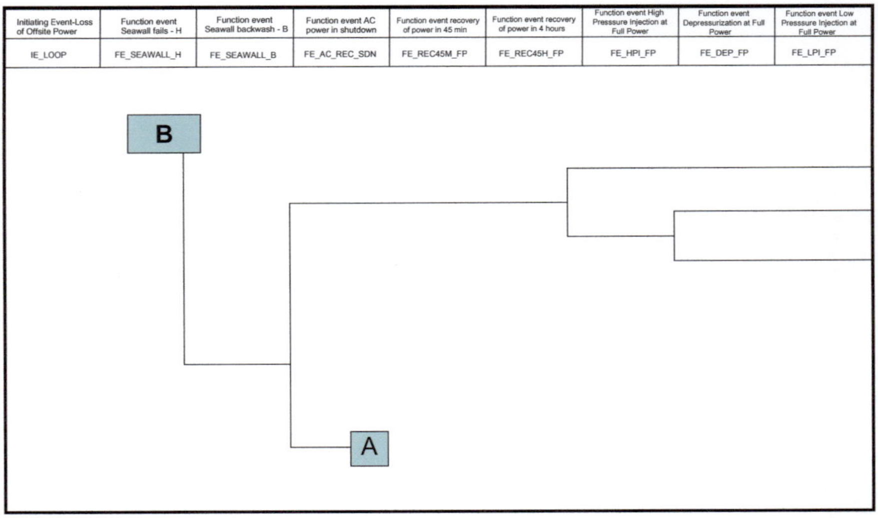

Fig. 2.9 Sample of an ET for Loss of Offsite Power (LOOP) connected to the Tsunami IE (2)

Table 2.6 Sample case of IE list with groups and sources identified [1]

Group of IE	IE group/Detailed list of IE covered by the group	Description	Power Levels		Support Documents	
			FP	SDN	DESIGN & SAFETY DOCUMENTS	OPERATING DOCUMENTS
RT	RT-_______-FP	Reactivity Transient at full power	×			×
RT	RT-_______-SDN	Reactivity Transient at partial power		×		×
LOOP	LOP-_______-FP	Loss of offsite power (LOOP) during full power operation	×		×	×
SBO	SBO-_______-FP	Station Blackout (loss of all offsite & internal plant power supply) at full power	×			×
SBO	SBO-_______-SDN	Station Blackout (loss of all offsite & internal plant power supply) at partial power		×	×	×
LPC	LPC-_______-FP	Loss of Plant Control (OCS) at full power	×			×

2.3 Databases

The Databases task of PSA starts specifically in the following:

- The Key Topic for the Database task ($\mathbf{KT}_3^{DB}$) is related to the need to have a specific database for an NPP.

  Example: Databases for a PSA are built based on information from

  - specialized standards/documents recognized in many other studies,
  - OPEX or from
  - databases from similar plants,
  - previous versions of PSA for the plant (if they exist). There are some representative groups requiring quantification, as, for instance:
  - Failure of NPP components (for which the boundaries are well defined). These could be active components (as, for instance, pumps, fans, etc.) or passive components (pipes, tanks, etc.).

    For each of them *a set of specific failure modes exists from previous studies and/or maybe defined by using techniques (for instance, Failure Mode Effects and Criticality Analysis)*;

    For passive components extensive databases exist for specific items (pipes, tanks, etc.) and special techniques (as, for instance, Markov chains is used to derive combined modes of failures or failures for passive systems if some conditions differ—as, for instance, another agent, gas instead of water is used).
  - Frequency values for IE, for which some specific aspects were illustrated in previous *Examples 1–3 solution* $\mathbf{S}_2^{\mathbf{PR}_2}$;
  - *Human Errors (HE)*;
  - *Probabilities assigned to decision point of high epistemic uncertainty, called Split Fraction*, which are basically a set of subjective allocated truth probabilities (from 0 to 1) to consider the degree of confidence in the decision made during PSA on a certain matter. Some specific issues related to Split Fractions were presented in Examples 1 and 2 of solution $\mathbf{S}_1^{\mathbf{PR}_1}$.

- Problem for the $\mathbf{KT}_3^{DB}$ ($\mathbf{PR}_3^{\mathbf{KT}_3}$) is to develop an overall comprehensive, traceable set of data for a given PSA study. Knowledge on the sources, rules to consider data from various databases, as well as a strategy to review and update the questionable inputs are essential for the accuracy of results. Database for a given PSA has, therefore, a set of high challenges due to uncertainties in the values, assumptions and differences between components, having diverse boundary conditions. Even if the goal is to have a plant-specific database, usually this goal is under continuous improvement, leading to the need to create mechanisms at each NPP for the development of its own database. However, most of the PSA studies have to cope with the best possible database to be used. Therefore, there are important challenges in choosing the data for a PSA. The diversity of the input to the data induces a need for the evaluation both of the impact of the chosen values and for the accuracy of the numeric results.

- Solution for the $\mathbf{PR}_3^{\mathbf{KT}_3}$ ($\mathbf{S}_3^{\mathbf{PR}_3}$) is to define rules to consider values of failure modes from the various databases if they differ and for various types of basic events.

Example 1 solution $\mathbf{S}_3^{\mathbf{PR}_3}$. Assuming that a certain *Failure Mode (FM)* is defined for a component (called further *Basic Event*) and that there is no plant-specific value for it, then a rule how to consider the existing data from other databases (for the same FM of a component with the same boundary conditions) has to be defined. In this case, instead of picking by a subjective decision of the existing values, the following approach is usually adopted:

- For each BE of a given database $BE_i^{DB_i}$ a confidence factor (w_i) on the value is allocated;
- *A weighted value* for the BE_i is then calculated as per formula (2.2)

$$BE_i = \frac{\sum_{i=1}^{n} w_i \cdot BE_i^{DB_i}}{\sum_{i=1}^{n} w_i} \tag{2.2}$$

Example 2 solution $\mathbf{S}_3^{\mathbf{PR}_3}$. The PSA database for a given study is composed of diverse types of data, as mentioned before in the description of the Key Topics of this task. A sample case of a PSA database in Table 2.7 illustrates the types of events expected:

- BE defined for the specific plant (if it exists) or as per Example 1 solution $\mathbf{S}_3^{\mathbf{PR}_3}$.
- CCF—Common Cause Failures for groups of BE.
- IE defined as per Examples 1–3 solution $\mathbf{S}_2^{\mathbf{PR}_2}$.
- SF defined as per Examples 1 and 2 of solution $\mathbf{S}_1^{\mathbf{PR}_1}$.
- HE—Human Errors of direct and recovery actions.

In most PSA studies, the assumed generic probability distributions for all BEs is log-normal. This is a generic approach even if the differences between various types of failures are known, as, for instance:

- Weibull for rotating components.
- Normal for most of the cases of components in which small numerous independent causes may lead to failures.
- Exponential for electronic components, etc.

For the log-normal distribution, the basic difference by comparison with the normal one is that many small random effects are, according to the central limit theorem, not additive like in the case of the normal distribution, but multiplicative. The physical meaning is that the various small causes are connected and conditioned that are additive for the normal distribution and multiplicative for the log-normal distribution (formula (2.3) and Fig. 2.10).

$$f(x, \mu, \sigma) = \frac{1}{x\sigma\sqrt{2\pi}} \cdot e^{-z^2/2} \quad x > 0 \tag{2.3}$$

where $z = (ln(x) - \mu)/\sigma$.

Table 2.7 Sample data for basic events, split fractions for seismic IE [1]

Event Probability	Type	Event	Description
4.08E-05	IE	Seism_S4	Seismic Initiating Event (ground acceleration 0.4 ÷ 0.6 g)
1.46E-04	IE	Seism_S3	Seismic Initiating Event (ground acceleration 0.3 ÷ 0.4 g)
5.00E-02	HE	MSSV-HE-REC	Operator fails to keep open at least one MSSV, after a seismic event (detailed analysis)
4.30E-02	HE	EPS.FAIL1H—HE	Operator fails to start EPS after a seismic event (detailed analysis)
4.63E-01	BE	Seism_BAT	Battery lost due to masonry partition walls collapse
7.66E-01	BE	Seism_S3_OFF.SITE.POWER	Loss of offsite power - fire induced
2.80E-02	SF	REC_Factor_HE	Recovery factor for dependent human actions (detailed analysis)
6.89E-01	BE	Seism_S2_CLSI-RLCH	Relay Chatter
1.80E-02	HE	ZHF-C2-Z01A	Recovery factor for dependent human actions (detailed analysis)
7.96E-01	SF	Split fraction factor _S3	Reduction factor for seism_S3 sequences
6.89E-01	BE	Seism_S3_RSW-RLCH	Relay Chatter
1 E-05	CCF	EPS CCF	EPS Common Cause Failures
1.80E-02	SF	Seism_S3_Rec_factor_HE	Recovery factor for dependent human actions (detailed analysis)

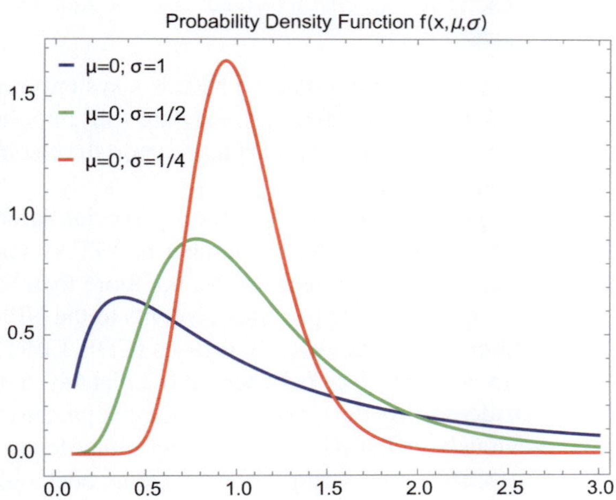

Fig. 2.10 The density for a continuous log-normal distributed variable

The log-normal distribution may be appropriate when uncertainties are known to be positively skewed. The implications are convoluted, because the quantification process, which is assuming a set of rules of calculating probabilities of scenarios (as it will be illustrated in the corresponding paragraph) assumes that they might be combined numerically easily at the level of medium values. However, if various distributions are assumed for different components, then a set of convolution integrals are needed and Monte Carlo approach may transform quantification in a very complicated task, without a significant gain in the global insights. Therefore, uncertainty analyses on the impact of using various distributions could be more pragmatic.

In conclusion, for each of the assumptions adopted during the FT task, connected with the database and quantification tasks, a set of possible sensitivity calculations is defined for the SUA task of the study and they are subject to PSA research activities, as the corresponding paragraph shows.

2.4 Event Trees

In a PSA study, the plant reaction to the challenges described by *IE is defined in scenarios developed by using Boolean logic.*

In a PSA, the NPP's reaction is described by following a set of principles:

1. The reaction is described *step by step*, usually assuming a time frame of analyses of 24, 48 and (rarely) 72 h. This time frame of the study is called Mission Time (MT);
2. In describing the reaction as per first principle, it is assumed that *No-miracle principle is applicable*, which means that if a certain system/component has to answer properly to the challenge, the study will assume that the probability *that it will fail to do so is not zero*;
3. The description is characterized at each step by the application of *triple 'S' approach*:

   - Systemic, assuming that the NPP is a system of systems. An NPP has around 200–300 systems. However, they are split into the following:
     - *Process Systems (PSy)*, which support the electricity production from nuclear fission;
     - *Support Systems (SupSY)*, for the Special Safety Systems, which are usually about 10% of the whole number of NPP systems;
     - *Special Safety Systems (SSY)*, not more than half number of the SupSy, that are designed to cope with accidents to the NPP, potentially leading to states described by various risk metrics (CDF, LERF, Total Risk).

     There is an interface between all the plant systems. For the PSA purpose, it is considered only the type of interface designed to cope with challenges (defined previously in Table 1.1 as the System Interdependence Matrix—SIM).
   - *Systematic*, considering all the systems declared by design and confirmed in operation *to be part of the reaction to a challenge.*

- *Structured*, which means that plant reactions in the format of Event Trees (ET) are described as in Figs. 2.1, 2.2 and 2.3.

4. The scenarios are developed considering the Boolean logic of Yes and No. A scenario assumes (Fig. 2.11) that each challenge will be coped with by the NPP barriers designed to respond gradually in time [1]. The resultant diagram is an Oriented Graph structure, with branches of Yes/No for the ET and branches of No for the FE. The FE, IE are described by probabilistic functions for their values. Usually, a PSA study *assumes a log-normal distribution*.

ET are, therefore, a set of logical binary combinations assuming that the components included in the reaction are of probabilistic type.

This is a very important aspect to be mentioned about ET in the PSA studies, which in many cases do not underline the fact that, the resultants scenarios leading to a certain damage state (described in risk metrics) are derived by using a Boolean type of binary states combinations and the probabilistic features are embedded only in the fact that the components of those scenarios are mainly of probabilistic type.

A timely description of the NPP systems answer (limited to the adopted Mission Time MT) is also assumed by design, i.e. reaction to the following:

- neutronic phenomena in seconds;
- thermal-hydraulic phenomena in minutes;

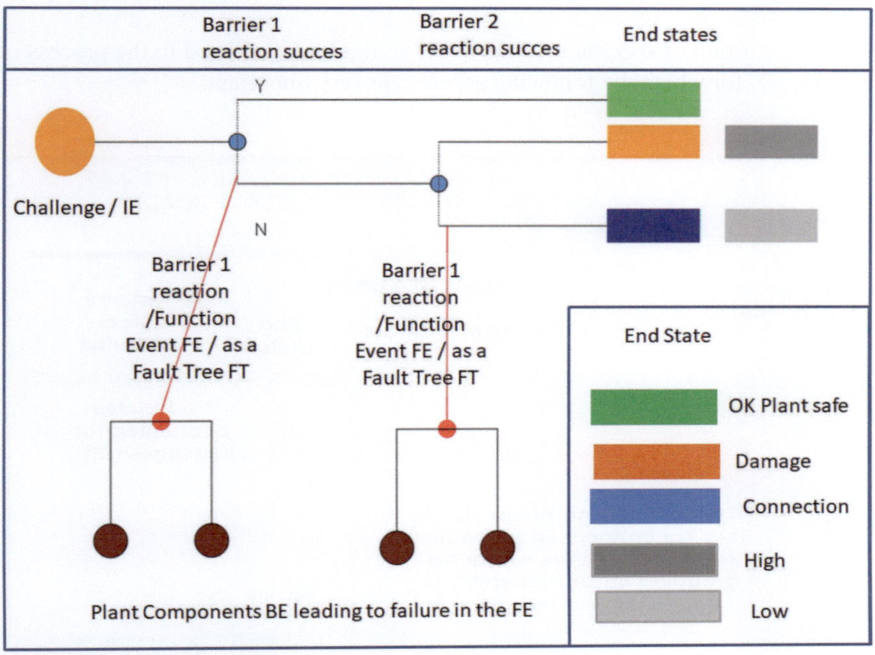

Fig. 2.11 The principle of building an ET (1)

- long-term combined neutronic—thermal-hydraulic—mechanic phenomena, also considering the operator, in hours.

This timely structured answer is reflected in the manner and the type of barriers, which are listed, that are mainly in three groups (in this order):

- First—reaction to reactivity effects;
- Second—reaction to cooling effects of the reactor;
- Third—reaction to the support of the first barriers and recovery.

The adoption of an MT value has an important impact, related to aspects as follows:

- Any descriptions of NPP reaction beyond the adopted MT value is not expected to be described in the PSA model;
- The model has to be changed for MT beyond the adopted MT.

The Fault Trees are integrated into the Event Trees as illustrated in Fig. 2.12 [1]. A PSA has to adopt one of the following approaches:

- Large ET and small FT to support FE, or
- Small ET and large FT.

There are advantages of each case debated in many PSA documents. However, for the purposes of this book, the important aspects are related to the problems that appear in building ET, which are, for most of the aspects, independent on the magnitude of ET.

Another group of specific features of ET in the PSA is related to the process of building ET, for which the following aspects are very important:

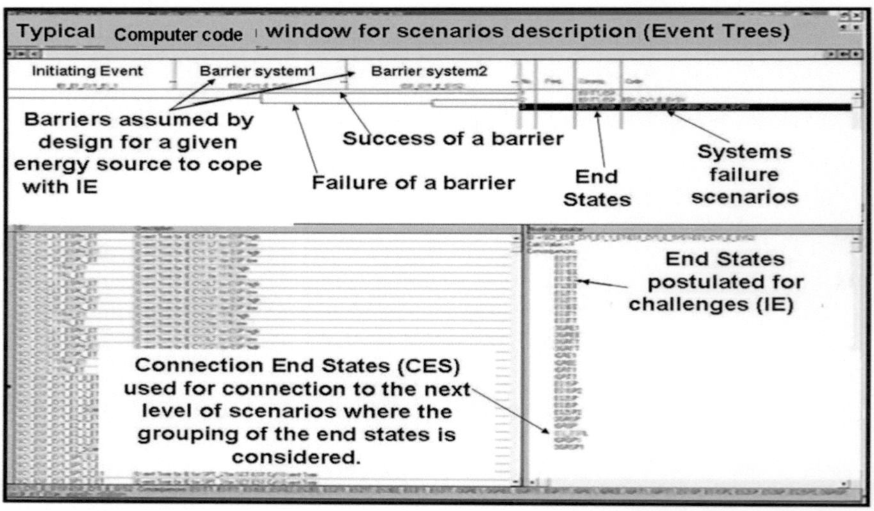

Fig. 2.12 The principle of building an ET (2)

- In order to properly define the FE, the Success Criteria (SC) for that barrier has to be defined, the main objectives of the SC tasks in a generic PSA are the following.
- The analysis in PSA has to start from existing deterministic (for instance, thermal-hydraulic) analyses simulating the course of accident progression. These analyses and assessments are supporting analyses for the success criteria formulation;
- The definition of the End States (ES) of a sequence has to start from the reactor core analysis and its status against risk metrics (CDF, LERF, Risk or no impact);
- The previous conditions are even more important for PSA level 2 (which considers the containment failure and might finish in a risk metric called LERF). The PSA level 2, ET are mostly based on the deterministic calculations and depends on the epistemic uncertainties of the codes providing the calculations;
- The safety-related functions defining SSY SupSy systems (Table 1.1) are, in terms of a PSA function, performed by them;
- for operator actions, SC are characterized by statements that certain actions are successfully carried out within a defined time window. There is a close connection between HRA (Human Reliability Analysis), systems analysis and SC formulation;
- the Internal IE PSA model is the starting point for the External events model;
- ET are defined in steps/hierarchy so that to be able to describe the behaviour of the reactor and systems, as well as the containment and associated type of applicable risk metrics;
- connection between ET is assured by using connectors which form, together with the switches at the level of components, the basis for further development of external events in an integrated PSA model for one unit or multiunits;
- the logical switches toolbox is used for internal/external events;
- an MT is adopted for the whole study;
- a first Key Topic for the Event Tree task ($\mathbf{KT}_4^{ET}$) is related to the definition of the risk metrics and their use in building ET for the NPP defined set of IE;
- *Problem for the* $\mathbf{KT}_4^{ET}$ ($\mathbf{PR}_4^{\mathbf{KT}_4}$): There are certain cases of NPP PSAs, in which one or all risk metrics are to be defined in a special manner. In this case, a new set of risk metrics has to be defined;
- *Solution for the* $\mathbf{PR}_4^{\mathbf{KT}_4}$ ($\mathbf{S}_4^{\mathbf{PR}_4}$): For the NPP for which the CDF does not have a meaning and the Release Categories (RC)—similar to LERF—and total NPP risk are evaluated as the NPP total risk metrics.

Example 1 solution $\mathbf{S}_4^{\mathbf{PR}_4}$: If an NPP of gas cooled type is challenged, then there will be no CDF, but various levels of releases (immediate and delayed) through the NPP building (Fig. 2.13) [2].

For the situation described in Fig. 2.9, a set of RC is defined (as illustrated in Figs. 2.14 and 2.15) [2].

ET are, therefore, in this case, similar to the PSA level 2 ET of water reactors, i.e. being focused on both failure and success paths, as it will be shown in the PSA level 2, next paragraph.

A second Key Topic for the Event Tree task ($\mathbf{KT}_5^{ET}$) is related to the approach needed to build an asset of ET in a triple 'S' overall NPP-integrated PSA model.

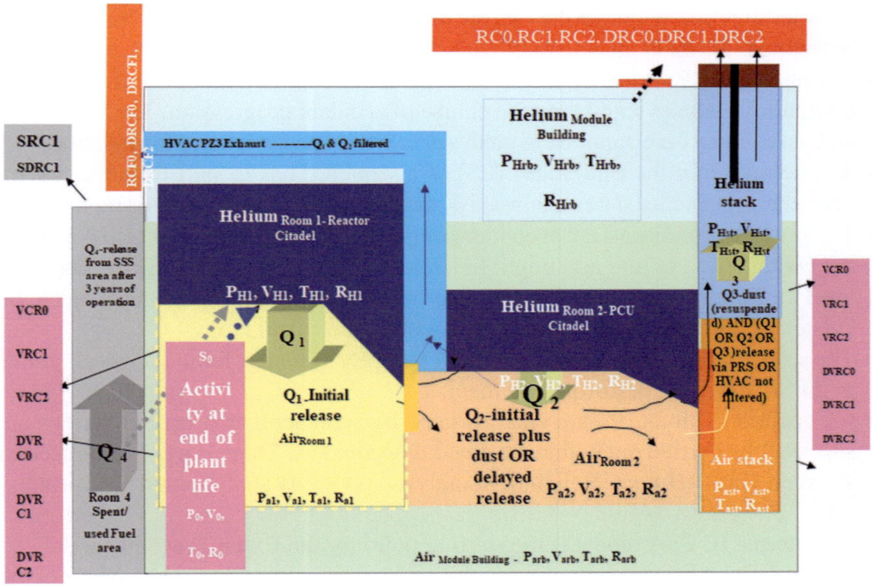

Fig. 2.13 Sample defining the end states, paths for releases and risk metrics in a gas- type reactor

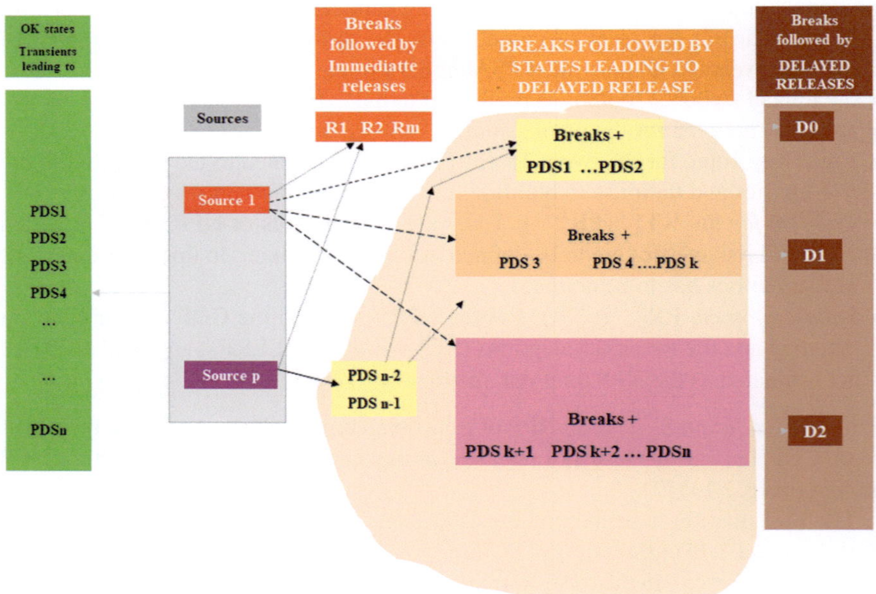

Fig. 2.14 Sample illustration of defining RC for a gas reactor NPP

Release category	Description	Release Category Frequency (RCF) Uncertainty Distribution (per Reactor - Year)				Basis for Risk Significance Criteria	Maximum increase in RCF considered insignificant
		Mean	5%	50%	95%		
RC I	No release, intact HPB	3.47	TBD	TBD	TBD	Covered by availability considerations	
RC II	Release of circulation activity only	3.50E-02	TBD	TBD	TBD	50% of RCF	2.00E-02
RC III -H	Delayed fuel release with pump-down and HVAC	3.70E-03	1.16E-04	9.39E-04	1.19E-02	50% of RCF	2.00E-03
RC III -N	Delayed fuel release with pump-down and no HVAC	1.85E-04	5.70E-06	4.69E-05	5.54E-04	50% of RCF	1.00E-04
RC IV -H	Delayed fuel release with HVAC	8.94E-04	3.27E-05	2.89E-04	3.39E-03	50% of RCF	5.00E-04
RC IV -N	Delayed fuel release with no HVAC	4.50E-05	1.59E-06	1.48E-05	1.69E-04	50% of RCF	2.00E-05
RC V - H	Delayed fuel release with oxidation, lift-off, HVAC	8.13E-06	4.78E-07	3.32E-06	2.59E-05	50% of RCF	4.00E-06
RC V - N	Delayed fuel release with oxidation, lift-off, no HVAC	9.91E-07	1.00E-07	4.73E-07	3.19E-06	100% of RCF	1.00E-06
RC VI	Loss of core and HPB structural integrity	3.80E-10	1.40E-11	1.45E-10	1.43E-09	100% of RCF	4.00E-10
	Total all analyzed sequences	3.51					

Fig. 2.15 Sample illustration of RC for a gas reactor NPP

The implementation of the triple 'S' approach for a generic PSA model starts from the description of an Integrated model based on the general considerations of PSA as a Complex Autopoietic Systems (CAS) (LP). CAS are systems, for which an autopoietic mechanism can be defined, leading to the system possibility not only to self-regulate, but also to recreate itself, as follows:

a. The system boundaries have to be clearly defined at any moment in time;
b. The system has to have components, being themselves CS;

c. The cause–effect law interactions have to be operable;
d. The system boundaries have to be self-produced by the system, as well as
e. The system components;
f. The rest of the components should be also be able for most of them to be self-produced by the system.

As a result of this general background, the PSA-integrated model is structured on three levels (as described by CAS in Figs. 2.16, 2.17 and 2.18) [3]:

- PSA level 1—CAS 1,
- PSA level 2—CAS 2,
- PSA level 3—CAS 3.

The problem for the $\mathbf{KT}_5^{ET}$ ($\mathbf{PR}_5^{\mathbf{KT}_5}$) is to define the tools to implement in specific computer codes the general type of connections and solutions described in and illustrated in Figs. 2.16, 2.17 and 2.18. They consist on defining detailed solutions for the implementation of the generic principles in practical PSA models. In order to reach this goal, the tools have to assume means on how to

- connect the Internal Events between them, so that to comply with the SIM and other input data by using optimal descriptors in the ET and FE/FT for the Internal IE PSA model (IPSA);
- connect IPSA model by new conditions to the External Events PSA (EPSA) without a large increase in number and dimension of ET and FT. Many of the details of such a set of tools are described also in the next task on Integration and quantification;

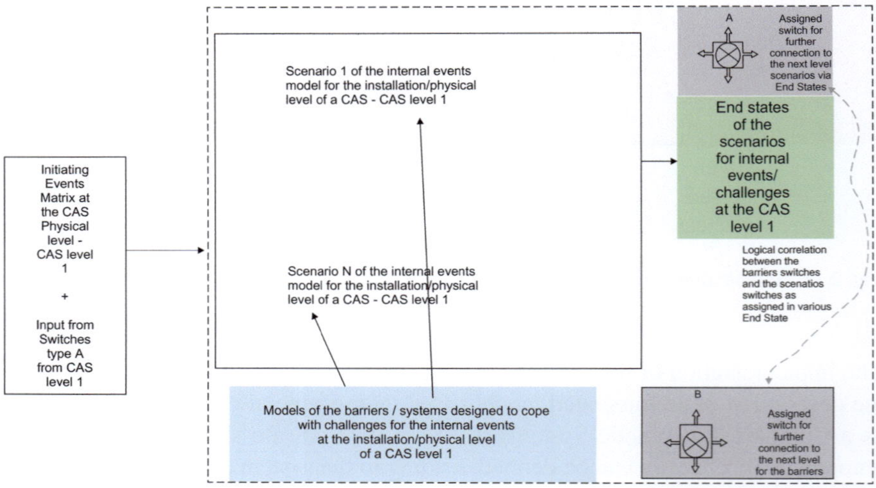

Fig. 2.16 Use of switches for ET in PSA level 1 for an NPP considered as a Complex System (CAS)

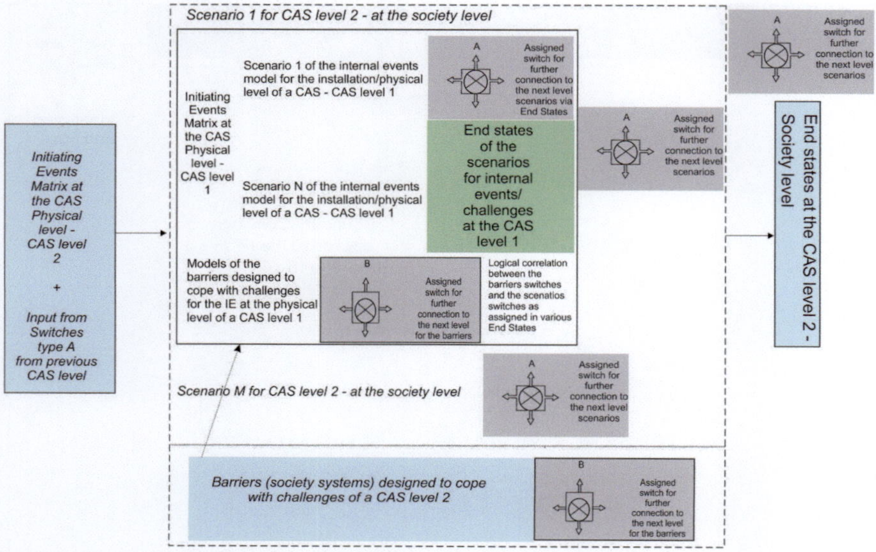

Fig. 2.17 Use of switches for ET in PSA level 2 for an NPP considered as a Complex System (CAS)

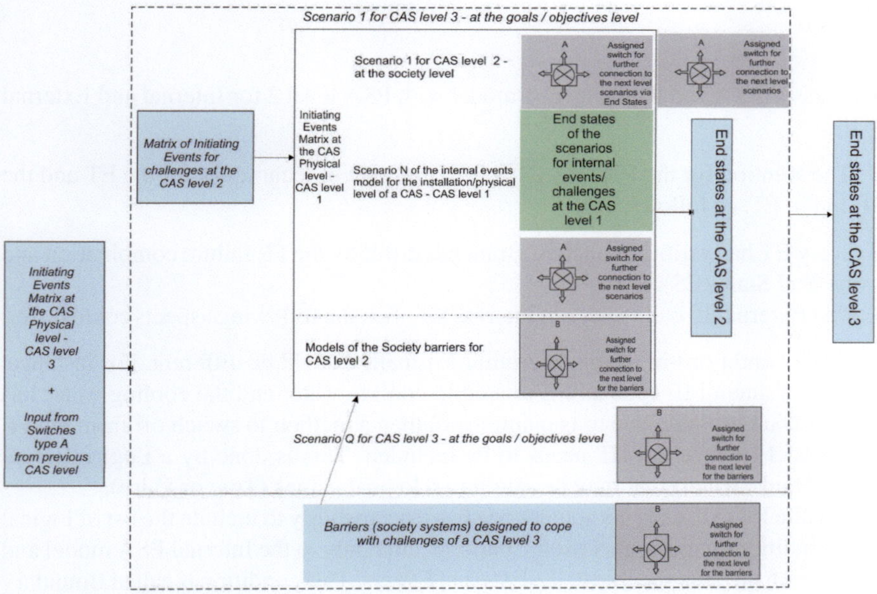

Fig. 2.18 Use of switches for ET in PSA level 3 for an NPP considered as a Complex System (CAS)

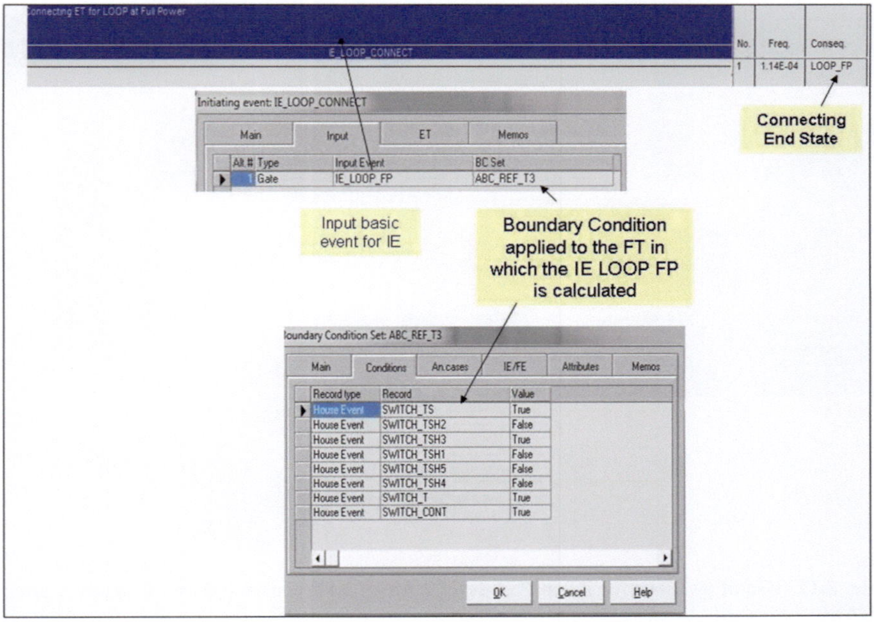

Fig. 2.19 Use of switches and BC for ET in a PSA software

- connect IPSA and EPSA level 1 model with PSA level 2 for Internal and External IE.

The solution for the $\mathbf{PR}_5^{\mathbf{KT}_5}$ ($\mathbf{S}_5^{\mathbf{PR}_5}$) is to use a set of techniques for the ET and the FE in the ET, as follows:

- Every ET has various branches. Branches differ by the FE failure combination and the End State (ES);
- If an External IE calls a certain Internal ET, then the following aspects could differ:
  - The conditions in which a certain FE is called could be different. For instance, if a Tsunami IE calls an FE related to cooling water and the cooling water has parts unavailable due to tsunami itself, then a method to switch off from the FE model for Internal ET needs to be included. This is done by a Logical Event (called Switch) that may be set on two logical values (True or False);
  - In the FE call, each computer code has the capability to include the list of logical conditions to disconnect some parts specific only to the Internal PSA model and switch on parts specific to the External Events. This condition is called Boundary Condition (BC);
  - The switch and BC tools are used also for different FE of diverse Internal IE in which the difference consists on which support system is needed (For different reactions, the FE might differ by availability or not of support systems like Instrument air, technical water, etc.);

– A detailed presentation of the use of Switches and BC is presented in the part of quantification.

BC and Switches. The set of logical conditions, mentioned under BC Example 2, solution $\mathbf{S}_5^{\mathbf{PR}_5}$. Figure 2.19 [1, 3] illustrates the set of logical conditions for ET in a PSA software [1, 4].

2.5 Fault Trees

FT are elements that describe the manner various systems or their parts are failing. As mentioned in the ET paragraph, some important generic features of the FT are to be mentioned:

- FT as oriented graphs using only 'NOT' logic. FT technique is used to define FE in the ET;
- However, FE are not FT as various FE may have common parts of the same FT and different parts of it.

There are other specific aspects of the modelling of the FT related to the pre-conditions to the task, which have a high impact on the PSA model as a whole, as follows:

- Specification of Boundary Conditions of System and of each component: The boundary of the assessment target system has to be specified to clarify the boundary between the system and other systems, as this aspect is very important to define the qualifications needed for it in case of internal events and external events (seismic, tsunami, etc.);
- Determination of Front Line Systems and Support Systems: If not only front line systems but also their support systems are required in order to ensure the function of the system, the boundaries between the front line systems and their support systems must be clarified, as well as their qualifications to various events;
- Specification of necessary operator actions.

The main objectives of the System Analysis that is the base for the FT description are, as follows:

- To identify and quantify the causes of failure for each plant system represented in the initiating event analysis and accident sequence analysis in such a way that for each safety function in accident sequence models, system models are developed with account for success criteria;
- System-level success criteria, mission times, time windows for operator actions, different initial system alignments and assumptions provide the basis for the system logic models as reflected in the model. A reasonably complete set of system failure and unavailability modes for each system is represented;

- Human errors and operator actions that could influence the system unavailability or the system's contribution to accident sequences are identified for development as part of the HRA element;
- Intersystem dependencies and intra-system dependencies including functional, human, phenomenological and common-cause failures that could influence system unavailability or the system's contribution to accident sequence frequencies are identified and accounted for.

The following aspects have to be considered in the modelling of the external events (seismic, tsunami, etc.) effects at the component level:

- The conditions of components designed to withstand the specific challenge (supports, watertight doors, etc.), that will have a great influence on the magnitude of damage inside buildings will be taken into consideration;
- In cases where damage to a component mentioned above causes a considerable increase in the amount of damage on systems and buildings, or multiple components, then dependency between fault trees must be properly dealt with;
- If target facilities have a correlation on damage (fragility) due to external events, as seen in components of the same type in the same section of a building, they could be modelled by using the base event of one of them.

It is also very important to mention that the results are expected to include combinations of both external events and random basic events from the internal model. The following aspects have to be considered as factors of functional loss/random failures:

- Outage because of component failure, testing or maintenance,
- Human error,
- Common cause failure classified as a dependent failure.

In particular, modelling of human error must be carried out properly by taking the following influences unique to tsunami events into consideration. For analysing human reliability in operator manipulations before and after the occurrence of a tsunami a validated HRA method has to be used. However, the highly stressful situation due to events like tsunami has to be able to be modelled by the adopted method.

Screening of Base Events. The number of base events may become enormous, so that some base events may be excluded from the quantification process on the basis of the concept of screening. Screening of base events will be carried out according to the following principles:

- If the damage probability of an assessment target component is very small for the top event, the base event will be regarded as an event that will not occur;
- In the case of a product event between a facility whose damage probability due to an IE is thought to be very high and a facility whose realistic yield strength against it is very high, then the ET scenarios induced by that IE will be considered.

A first Key Topic for the Fault Tree task ($\mathbf{KT}_6^{FT}$) is actually to build correctly an FT.

Problem for the $\mathbf{KT}_6^{FT}$ ($\mathbf{PR}_6\,\mathbf{KT}_6$) is to prevent the appearance of a common mistake in building FT, consisting on not following the three main principles of PSA mentioned in the previous paragraph on generic PSA rules:

- Step by step,
- No-miracle,
- Triple 'S' approach.

Solution for the $\mathbf{PR}_6\,\mathbf{KT}_6$($\mathbf{S}_7\,\mathbf{PR}_6$) is combined with the fact that the starting point and the process of FT construction do not follow some strict rules, resulting from application of the generic features presented before in this paragraph.

Example 1 solution $\mathbf{S}_6^{\mathbf{PR}_6}$: Illustrate in more detail how to apply the principles stated before for a specific case. The following steps are to be followed:

- If the system represented in Fig. 2.20 is one assumed to be called by a specific FE in an ET, then the most important starting point is to define the function that it has to perform. Asking the proper question will define the main question (called the TOP of the FT). In this case, it may be 'Do we have flow in point B when required and in the conditions from the FE?'
- From the FT TOP, a series of questions of what might go wrong to get to it are asked. However, the questions follow the system diagram and its presentation in a special format called (Reliability Equivalent Diagram (RED)—as represented in Fig. 2.20). Therefore, one might ask in the following order, the questions related to the fact is there is a flow after:
- RV and if not which were the causes?
- V1 and if not which are the causes?

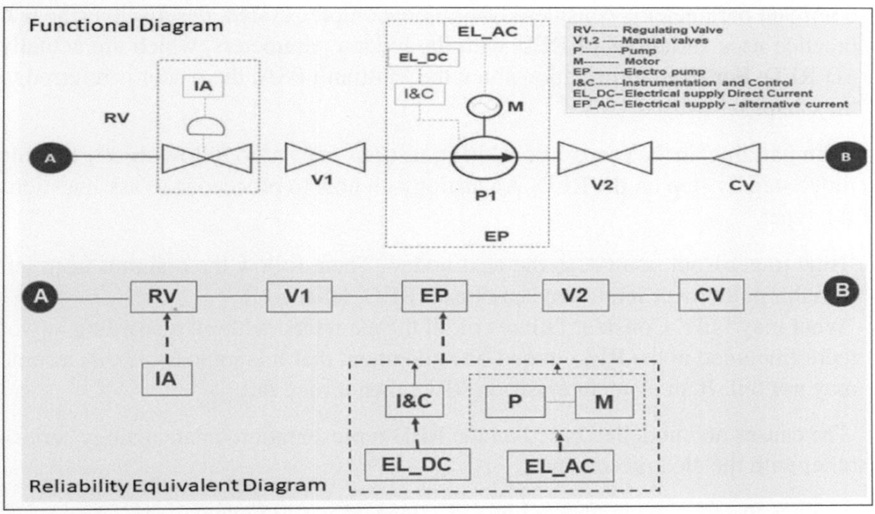

Fig. 2.20 Building a reliability equivalent diagram (2D) starting from a functional diagram

- EP and if not which were the causes?
- V2 and if not which are the causes?
- CV to point B and if not which are the causes?

The RED as used in the present and as in the sample from Fig. 2.20 represents the system assumed to be a system of subsystems reacting to a challenge. The possible reaction that can lead to a failure of the system is built up in a time sequence logic and considers the functional connections between the subsystems/components of the system. However, there are two aspects to be mentioned:

- The subsystems might be dependent on one each other (for instance, IA may depend on DC power which in turn depends on other support subsystems/ components). Therefore, the usual case is that the model leads to a system that was previously considered as a supporting one. In such cases, the search for the primary cause has to be stopped and it has to be decided which of the failures for the support components/subsystems was the first at all. This is decided by knowing the time description of the system reaction and the decision of stopping the search for the primary cause (called 'Breaking the Logic Loops') is taken;
- Another important aspect is related to the fact that the subsystems/components are assumed as elements failing or not as a function of time for ONE dominant aspect. However, there are various parameters that could have a major impact on the subsystems and systems failures (except the dominant ones) and these constitute a set of parameters. Of all the parameters there are some neglected ones, which need to be reconsidered in new RED versions. Actually, in such case the RED will depend on one variable (the dominant failure mechanism) and a hidden (parametrically to be considered) one. Therefore, the RED model is not bidimensional (2D-RED, as in Fig. 2.20), but three-dimensional (if the hidden dominant parameter is considered) for more complex system descriptions the real practice uses variations of RED with the hidden parameters, which are actually 3D-RED. For more information about the Multiunit PSA, the reader is referred to the Chap. 3.

Summarizing, in the process of asking questions one has to follow strictly the rule to move step by step on the RED. An analogy on how to proceed is to ask questions of the type:

- 'How to get from point A to the next node?' Then, follow the diagram from left to right, point by point, to the last one on RED, labelled B;
- 'What may fail?' Consider failures of all the elements, without discarding any of them (included in the RED) based on a judgment that it is impossible that a some may not fail. If an element exists on RED, then it may fail.

The causes are modelled based on the RED representation, in an analogy series–parallel with the electric schemes:

- the lines in series are connected by OR connectors (OR Gates),
- while the parallel connections are modelled between them as AND nods (AND GATES).

The logic combination of the OR and AND gates leads to a set of failures that could describe the TOP. In the case from Fig. 2.19, the TOP is described by the equation:

$$TOP = RV\dot{+}V1\dot{+}EP\dot{+}V2\dot{+}CP \tag{2.4}$$

The Boolean logic decision points, called 'Nodes' in graph analogy and 'Gates' in FT description, are mainly the following:

OR	TRUE if at least one input event is TRUE
AND	TRUE if all input events are TRUE
K-of-N (K/N)	TRUE if at least K of the N input events are TRUE
NOR (NOT OR)	TRUE if none of the input events TRUE (all input events FALSE)
NAND(NOT AND)	TRUE if not all input events TRUE (At least one input event FALSE)
XOR (Exclusive-OR)	TRUE if an odd number of its events are TRUE, and FALSE otherwise
Switch	Logic value of TRUE or FALSE.

For the Boolean expression 2.4, the FT calculation is also performing the quantification by calculating the probabilities of the gates, as in formula 2.5:

$$Q_{TOP} = Q_{RV} + P_{RV}Q_{V1} + P_{RV}P_{V1}Q_{EP} + P_{RV}P_{V1}P_{EP}Q_{V2} + P_{RV}P_{V1}P_{EP}P_{V2}Q_{CV} \tag{2.5}$$

where $P_i = 1 - Q_i,\ i \in \{CV, EP, RV, V1, V2\}$.

The computer codes are using Boolean logic rules by performing a TOP-Down calculation of the critical paths to TOP and a Bottom-UP verification. As a result, a set of critical minimal paths to the TOP are confirmed (Minimal Cut Sets—MCS).

The quantification of MCS is performed in the computer codes with various approaches. The best known based on the classic oriented graphs modules in all the existing codes are based on calculating MCS approximations:

- The rare event approximation, which is usually the normal first-order approximation is used as a good approximation for the cases when probabilities are low. In this case, the TOP event probability is the sum of the unavailabilities of the MCS (formula 2.6):

$$Q_{TOP} = \sum_{i=1}^{n} Q_{MCS}(i) \tag{2.6}$$

- Other improved approximations for rare events, as, for instance, the min cut upper bound (Formula 2.7): somewhat better approximation than the rare event approximation. The min cut upper bound formula is as follows:

$$Q_{TOP}^{MCUB} = 1 - \prod_{i=1}^{n}(1 - Q_{MCS}(i)) \tag{2.7}$$

$Q_{MCS}(i)$ is the unavailability of MCS_i. As mentioned before, all the MCS calculations in a PSA are based on the results from the *Database task* and represent *mean values* of the unavailabilities that follow certain assumed distributions, such as Beta distribution, log-normal distribution, Beta-binomial distribution, etc.; in most PSA, the assumed distribution is log-normal.

A second Key Topic for the Fault Tree task ($\mathbf{KT}_7^{FT}$) is to optimize the number of FT to support FE in all the ET of a PSA model.

- *Problem for the* $\mathbf{KT}_7^{FT}$ ($\mathbf{PR}_7^{\mathbf{KT}_7}$): A mature PSA may have 50 ET for the internal events and at least as many for each area and external events, if these are not built in an integrated model IPSA and EPSA;
- *Solution for the* $\mathbf{KT}_7^{FT}$ ($\mathbf{PR}_7^{\mathbf{KT}_7}$) is to assume if special modelling actions are not taken, then the magnitude of the PSA model becomes hardly manageable, not talking about the computer modelling issues. Due to this situation, it is very important to use another important aspect from the Boolean toolbox of the PSA methodology: *Logical conditions and equations*.

Example 2 solution $\mathbf{S}_7^{\mathbf{PR}_7}$: The solution to solving this problem is to use logical events (called House Events or Switches). In Figs. 2.21 and 2.22, it is shown how such switches are used. The initial switch introduction action starts with the existence of only one basic event $ALTHEAT\_N$. In order to introduce a logic combination of how to assure the existence of this event and of the tsunami basic

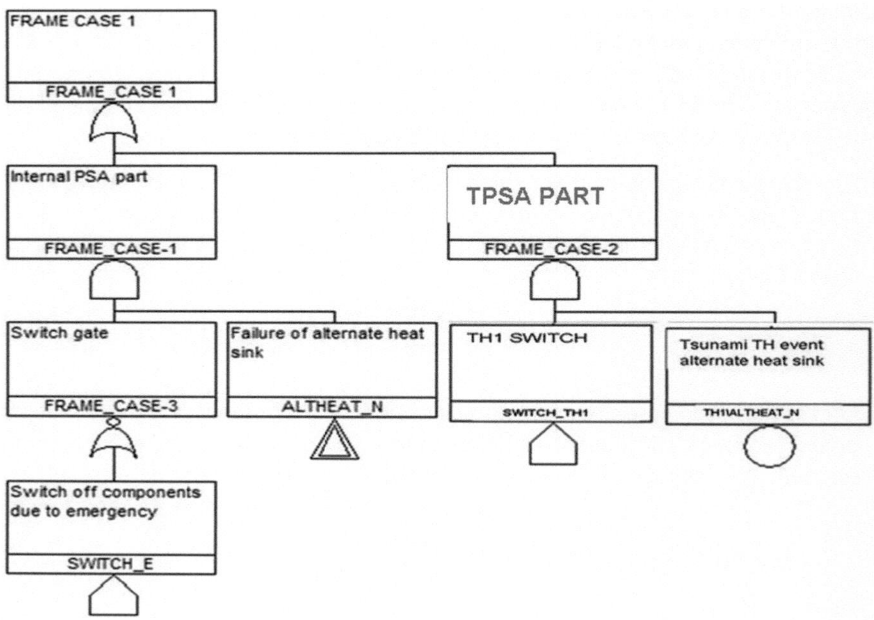

Fig. 2.21 Use of Switches (House Events) for IPSA and EPSA

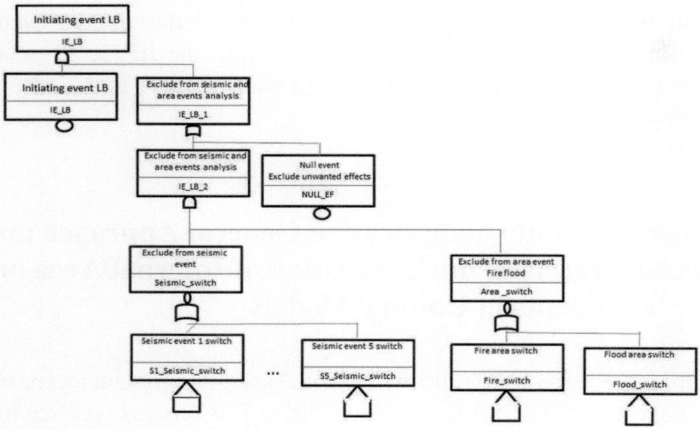

Fig. 2.22 Use of switches for area and external events in IE FT

event ($TH1\backslash ALTHEAT\_N$), in case that $IE\ TH1$ has to be considered, is done as follows:

- Considering the Tsunami IM (Table 2.3) between IE from IPSA model triggered by Tsunami the FE for the new Internal FE (existing both in IPSA and EPSA), a certain FE (coded as FRAME-CASE 1) is developed to cover both cases (IPSA and EPSA). For this purpose, logical switches ($Switch\_E$ and $Switch\_TH1$) are used (Fig. 2.21).
- Under the gate FRAME CASE 1 (as being the place, where there is the first highest OR gate above the internal basic event), a module to consider switches for including external events are inserted; two logic modules are introduced as in Fig. 2.21:
  - one for the internal basic event (on the left side),
  - one for the Tsunami initiator $Th1$ (on the right side).
- The events coded as $Switch\_E$ and $Switch\_TH1$ are logical-valued variables (TRUE or FALSE);
- In Fig. 2.21, the normal initial status of those logical events is FALSE [1, 5]. The effect will be as follows:
  - if both switches are normal, then $IPSA\ BE\ ALTHEAT\_N$ will be enabled, then the $IPSA\ BE$ will be valid. This happens because a NOR gate of a FALSE event will lead to a TRUE one and the gate; therefore FRAME CASE 3 will be valid TRUE and calculated and the gate FRAME CASE 2 will be FALSE and not calculated; the tsunami $BE(TH1\backslash ALTHEAT\_N)$ will be excluded;
  - If $SWITCH\_E$ is FALSE and $SWITCH\_TH11$ are TRUE, then both internal and tsunami basic events are calculated.
- As mentioned in Example 2 solution $\mathbf{S}_4^{\mathbf{PR}_4}$, various combinations of more than one switch can be used and defined as a logic rule of BC.

A similar process is followed for more than one external/area events (as in Fig. 2.22) in a special type of FT, the FT generating the IE [1, 5]. In this case, more switches (for fire, flood, seismic) might be used in on FT to generate an FT for the calculation of the IE.

2.6 Integration and Quantification General Approach and Special Aspects of the Integration of Internal/Area or External Events in Unitary Models

After developing ET and FE/FT, for which some specific problems were mentioned before, the next step to the PSA is to connect them. This process is called Integration of FT into the ET. The ET are connected with FE called (Fig. 2.23) and the result is a set of combined scenarios leading to various End States (ES) [3], which are of various types:

- with no impact on risk (OK ES);
- connecting ES to other ET;
- ES with impact on risk for which CDF is calculated.

The ES leading to core damage are evaluated by the CDF risk metrics.

From the graph modelling point of view, the integration is a combination of two types of oriented directed graphs; into the ET, in the branches with NO nodes, the FT are connected by calling their TOP gates. The resultant combination is a set of branches defining the combination of failures that could lead to the core damage and plant risk (for PSA level 1 IPSA and EPSA, this is core damage and the quantification is CDF). This process is illustrated in Fig. 2.24 [3].

As shown in the introduction paragraph, PSA has various important tasks, connected between them. If considering that each task produces a set of states from the PSA like NPP description, then the whole description of NPP by using the PSA

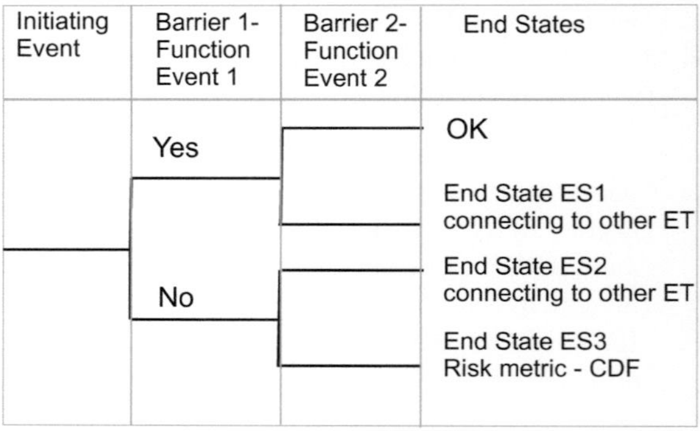

Fig. 2.23 ET schematic representation

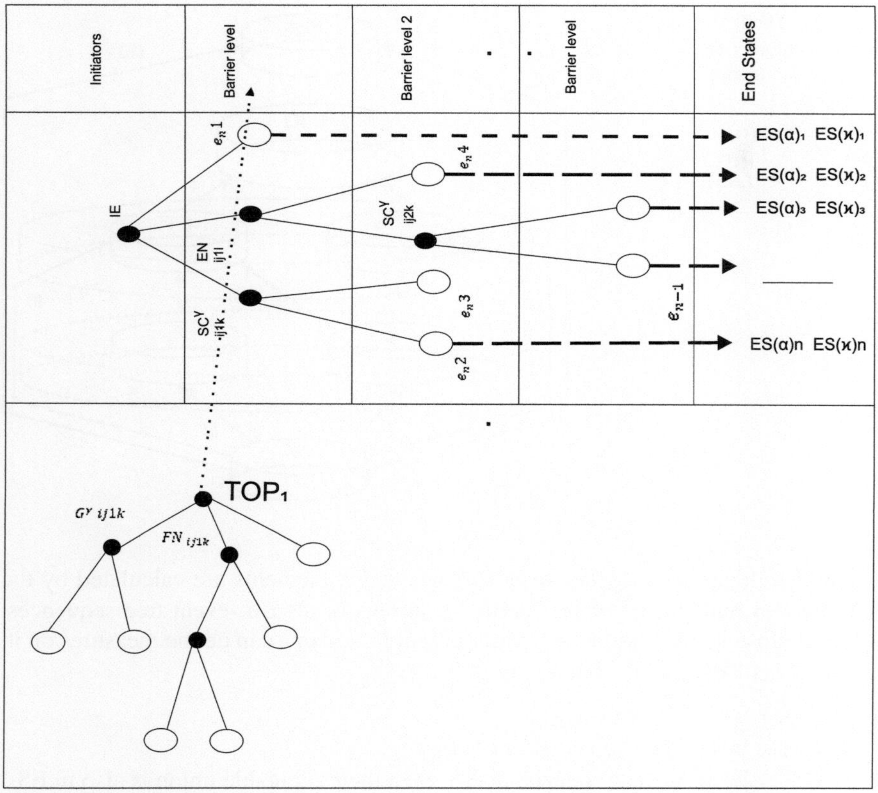

Fig. 2.24 Illustration of the integration process of FT into the FE, as defined in the ET

approach might lead to the generation of an algebraic structure. The interfering tasks generating such an algebraic structure are illustrated in Fig. 2.25 [3].

The construction of ET is performed in such a manner to be able to build an integrated PSA model, by assuring combination, in a 'matrioshka type' of approach:

- Internal model,
- Area events (flood, fire),
- External events,
- Multiunit model,
- Multisource model.

Therefore, if considering the PSA model as a complex system, that generates an algebraic structure by modelling an NPP, then the measure of this structure is called risk (with various forms, depending on PSA level: CDF for level 1, LERF for level 2 and Risk for level 3). A simplified generic representation for the risk metrics evaluations is given by formula (1.1).

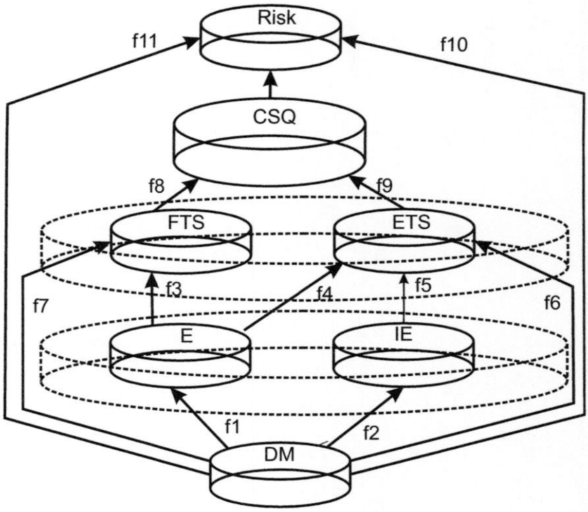

Fig. 2.25 The PSA tasks and their interaction to generate an algebraic structure: $f1 = f_{ed}$; $f2 = f_{ied}$; $f3 = f_{efts}$; $f4 = f_{eets}$; $f5 = f_{ieets}$; $f6 = f_{dmets}$; $f7 = f_{dmfts}$; $f8 = f_{ftscsq}$; $f9 = f_{etscq}$; $f10 = f_{dmcsq}$; $f11 = f_{dmr}$

The ES define a σ-*algebra* over ES, where ES elements are calculated by the function ES and consist of the family of subsets of all ES_i-event tree sequences, which is closed under countable set of operations and one can define measures on it, where ES is a σ-*algebra* if and only if:

1. The empty set of ES_i is in ES;
2. If ES_j is in ES then so is the non ES_j;
3. If ES_1, ES_2, ES_3 is a sequence in ES, then their countable union is also in ES.

From pragmatic point of view, the modelling of PSA as an algebraic structure actually leads to a set of

- matrices describing plant reaction,
- vectors describing the challenges (IE),
- vectors defining limitations due to various epistemic limitations and assumptions, as shown in Fig. 2.26 and Eq. (2.11) [1, 6].

$$RM_1^1 = CDF^1 = 1 - Sq\_cdf_1^1 \cdot Sq\_cdf_2^1 \cdot Sq\_cdf_3^1 \cdots Sq\_cdf_n^1 \tag{2.8}$$

$$RM_1^2 = CDF^2 = 1 - Sq\_cdf_1^2 \cdot Sq\_cdf_2^2 \cdot Sq\_cdf_3^2 \cdots Sq\_cdf_n^2 \tag{2.9}$$

$$RM_1^m = CDF^m = 1 - Sq\_cdf_1^m \cdot Sq\_cdf_2^m \cdot Sq\_cdf_3^m \cdots Sq\_cdf_n^m \tag{2.10}$$

$$RM_1^{ALL} = CDF = \sum_{i=1}^{m} CDF^i \tag{2.11}$$

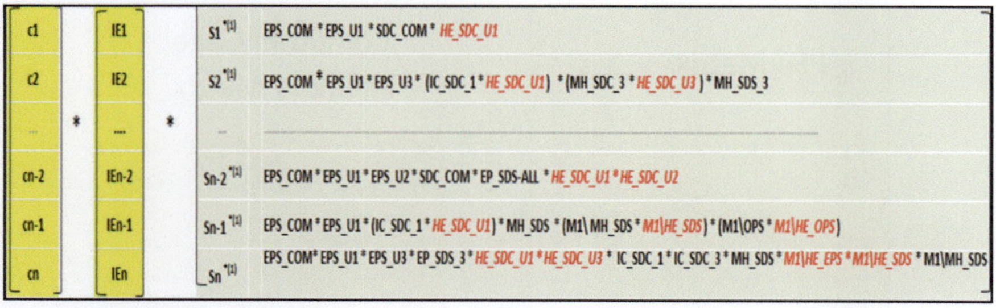

Fig. 2.26 Sample representation of the PSA as a process of building an algebraic structure: 1

$$\begin{bmatrix} IE_1 \\ IE_2 \\ IE_3 \\ \vdots \\ IE_n \end{bmatrix} \otimes \begin{bmatrix} c_1 \\ c_2 \\ c_3 \\ \vdots \\ c_n \end{bmatrix} \otimes \begin{bmatrix} a_{11} & \dots & a_{1n} \\ a_{21} & \dots & a_{2n} \\ \vdots & \dots & \vdots \\ a_{n1} & \dots & a_{nn} \end{bmatrix} = \begin{bmatrix} S_1^1 \\ S_1^2 \\ S_3^1 \\ \vdots \\ S_n^1 \end{bmatrix} \tag{2.12}$$

The PSA result is also represented by available algebraic tools. This is a very important aspect, as it is related to the

- steps of building PSA model and derivation of the risk metrics;
- the use of computer codes to manage very large models in a format of matrices and vectors, which actually are the PSA model itself.

In order to perform the support for PSA tasks, a full similitude between the PSA model and the computer code memory management is being built (Fig. 2.27). Understanding these aspects is a very important step in improving and optimizing PSA models. A very important part of those codes is related to the modules defining the tables of the interface between the places of a certain element in the PSA structure versus its place in the code memory.

The result of the integration process consists of a set of values for the occurrence frequency of accident sequences and for the Core Damage Frequency (CDF). These values are to be also evaluated with their uncertainty results. In the quantification and results evaluation phase, it is very important to perform the evaluation of the recovery actions.

However, special analyses are needed to evaluate the impact on the results of some aspects for which PSA models have limited tools (e.g. CCF for multiunit or HRA for external events).

For instance, for the HRA model after external events, there are important operator-related aspects to be considered, as the operator recovery actions cannot be credited if some conditions are not fulfilled, as follows:

- Operators should be in a safe situation after the external event in order to be able to perform recovery actions;
- The NPP-affected parts have to be accessible;

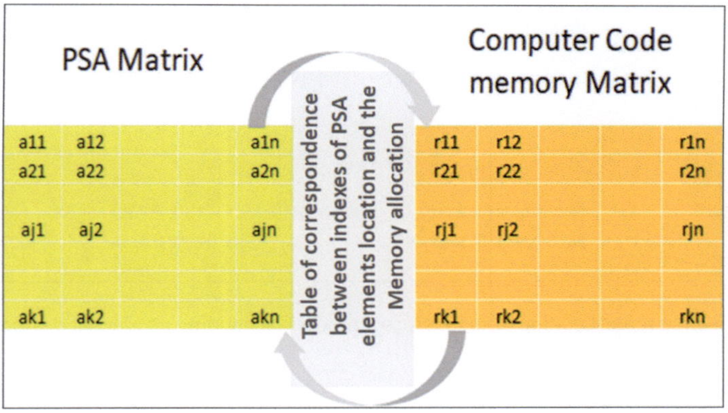

Fig. 2.27 Similitude between PSA model and PSA computer codes structures

- The assumed time action is considered, etc.

Related to the integration task, it is also important to mention the dilemma Small ET versus Large FT or Large ET versus Small FT dilemma specific to the PSA tasks. These two approaches are different from each other in the following way:

- The targets of the small event tree method do not include the support systems that will implement relaxation functions, whereas
- those of the large tree event method include both front line and support systems that will implement relaxation functions.

In the case of the Small ET versus *Large FT*, the ET are composed of small event trees method and the large event trees. This method creates an ET by using barriers (Special safety Systems —SSy, Support Systems—SupSy, etc.) to assure plant reaction on various events, including immediate operator and recovery actions. The operating conditions of support systems that enable front line systems to operate appropriately will be taken into consideration in individual headings separately (FE), in principle. Buildings, structures and components having a barrier role for a given event might be simultaneously damaged, when an event reaches a plant. The small event tree method is a simple direct presentation on the impact on the plant of an event, but it creates a much more number of FE.

In the case of the Large ET versus *Small FT*, the ET are created by using both front-line and support systems to implement barrier/safety functions approach and the conditions of facilities or manipulations by operators that are important for the development of accident sequences, etc., as its target headings. Event trees for front-line systems and those for support systems will be created separately, and then, both event trees will be connected to form an event tree describing their corresponding event. The large event tree method clearly shows the dependency between front line systems and support systems and thus makes it possible to easily identify the damage conditions of buildings, structures and components that will influence multiple safety functions simultaneously. However, the arrangement of such event trees will become complicated.

It is important to mention that the final Minimal Cut Sets (MCS) of the sequences have to be compliant in both approaches and that the difference is mainly a question of technique and depends on the existing information and goals of the tasks and PSA in general.

In any approach, the risk metrics for PSA is calculated in a code evaluation approach as per formulas (2.6) or (2.7). However, the generic formula for risk metrics, which is represented in Fig. 2.26 is

$$CDF = \sum_{j=1}^{m}\left(1 - \prod_{i=1}^{n}(1 - Q_{MCS}(i, j))\right) \tag{2.13}$$

where i is counting the components in a given TOP of a FT, while j is counting the sequences $Seq(i, j)$ from Figs. 2.26 and formulas (2.8)–(2.11) which lead to risk state.

In any approach adopted for the type of combinations ET–FT, the optimization process and the necessity to keep the process auditable will lead to the use of some special techniques (having the same goal in any computer code, even if with different practical implementation rules), as follows:

- Evaluation of the assumptions and their qualification in qualitative or quantitative manner, as, for instance, Split Fractions (SF) on decisions moments, which are formalized in subjective probabilities on the credibility of decision in the FT, FE or ET;
- ES in the form of markers to assure connection between various parts of the PSA model, used in the ET (Connecting ES);
- Logical equations in order to define parts of the PSA logic available under certain conditions:
  - Switches (House Events) in the FT;
  - BC in the ET, including a combination of various Switches.

The use of all techniques mentioned above actually supports the implementation of the generic principles presented previously in the Figs. 2.16, 2.17 and 2.18 in the PSA model. This process has the following steps:

1. Define Connecting Event Trees (Connect ET) of the PSA model and their associated FE/FT;
2. Develop ET to describe NPP reaction to IE, for various cases of IPSA and EPSA;
3. Describe the containment reaction to IE for PSA level 2;
4. Adapt the use of the PSA model for applications.

The first step is to build Connect ET (Fig. 2.28) of the PSA model and their associated FE/FT. The connecting ET are used to build the Aggregate IE event part for the I_IPSA_EPSA (PSA-integrated model):

- Building the IPSA model: Development of the IPSA starts with the evaluation of the Plant and challenges to it, previous studies and events and building of the list of IE for internal events;
- There are various possible ways to build the ET in order to assure possible future connections of the external PSA type to the IPSA. In the example from Figs. 2.28, 2.29, 2.30 and 2.31, the method adopted starts from the IE list and for each IE two sets of ET are built:
  - One ET used to assure the connection between the IE defined in the databases or as a result of building special FT for the calculation of IE. This type of ET is illustrated in Figure 36 (called Connecting ET). The ES of these ET are defined as consequences, with the same main code name as for the IE considered (TRAN for IE transient, LOOP for IE Loss of Offsite Power, etc.);
  - Another type of ET (as illustrated in Figs. 2.29, 2.30 and 2.31) is related to its usual description as the plant reaction for each IE type.
- These ET have as input the consequences defined by the first category (Connect ET);
- The ES are related to the risk metric under consideration (the main runs and models in this example are related to CDF or LERF but developments are available as shown in Fig. 2.28 for PSA level 1 and Figs. 2.30 and 2.31 for PSA level 2 (for an NPP with one circuit) [1].

The Key Topic for the Integration and quantification ($\mathbf{KT}_8^{IQ}$) is related to the magnitude of the model and the need to manage Integrated $IPSA - EPSA$ (I_IPSA_EPSA) models.

The problem for the $\mathbf{KT}_8^{IQ}$ ($\mathbf{PR}_8^{\mathbf{KT}_8}$) is how to optimize the number of FT and FE and the memory required for their management in an (I_IPSA_EPSA) model.

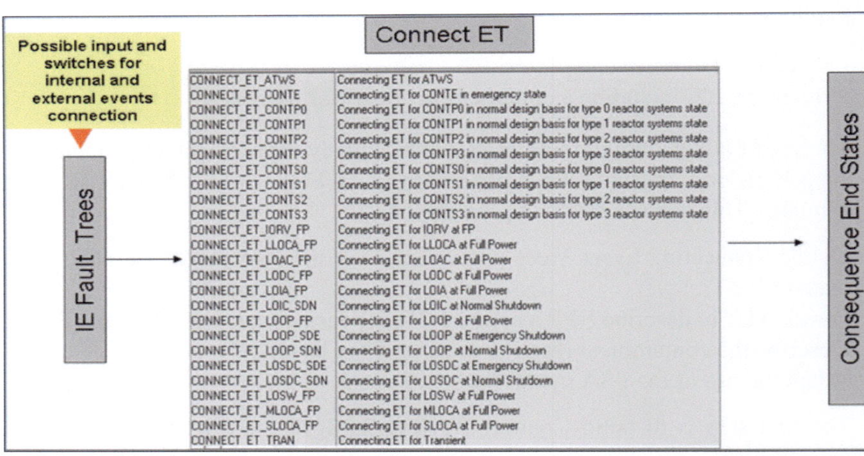

Fig. 2.28 IPSA model—list of connecting ET

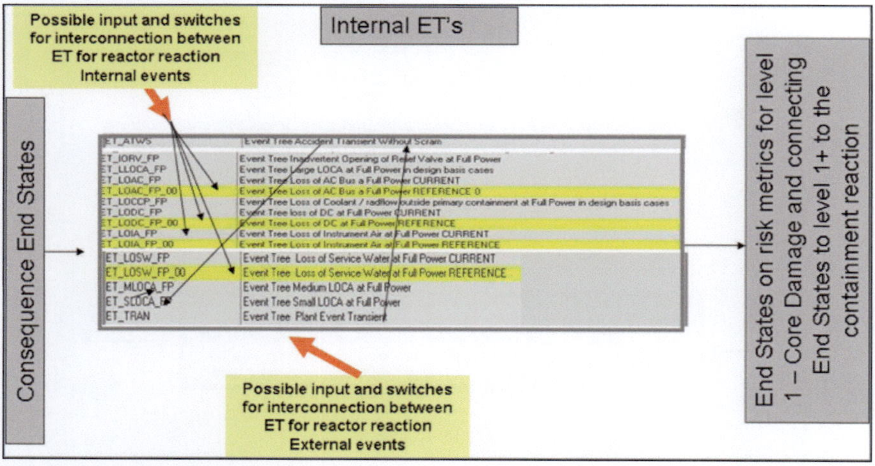

Fig. 2.29 IPSA model—building of the ET themselves

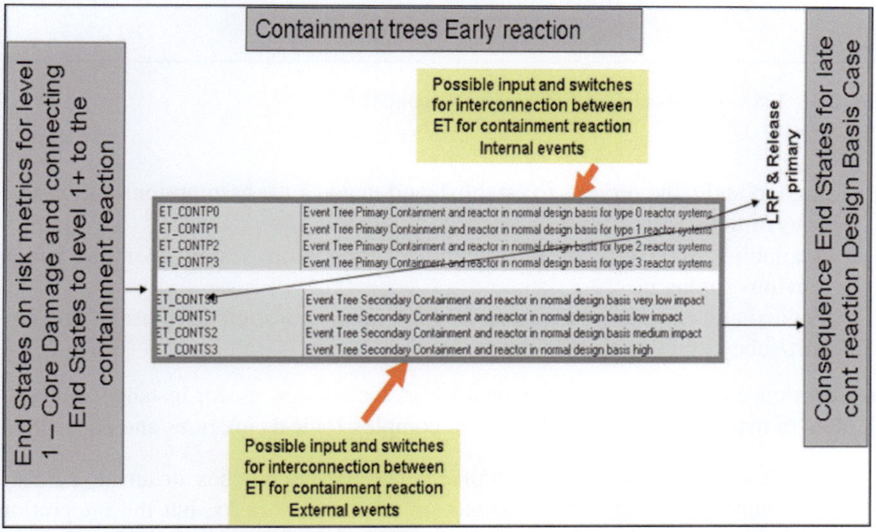

Fig. 2.30 IPSA model—building of the containment ET: 1

The solution for the $\mathbf{PR}_8^{\mathbf{KT}_8}$ ($\mathbf{S}_8^{\mathbf{PR}_8}$) is to use approaches practically applicable to ET and FT, as mentioned in the general presentation of the integration task.

Example 1 solution $\mathbf{S}_8^{PR-8}$: The optimization of the (I_IPSA_EPSA) model depends on a series of actions taken before, during and after the study is completed. They consist of the following:

- Managerial and general project approaches of the study, as follows:
  - A clear definition of the PSA objectives and its intended use;

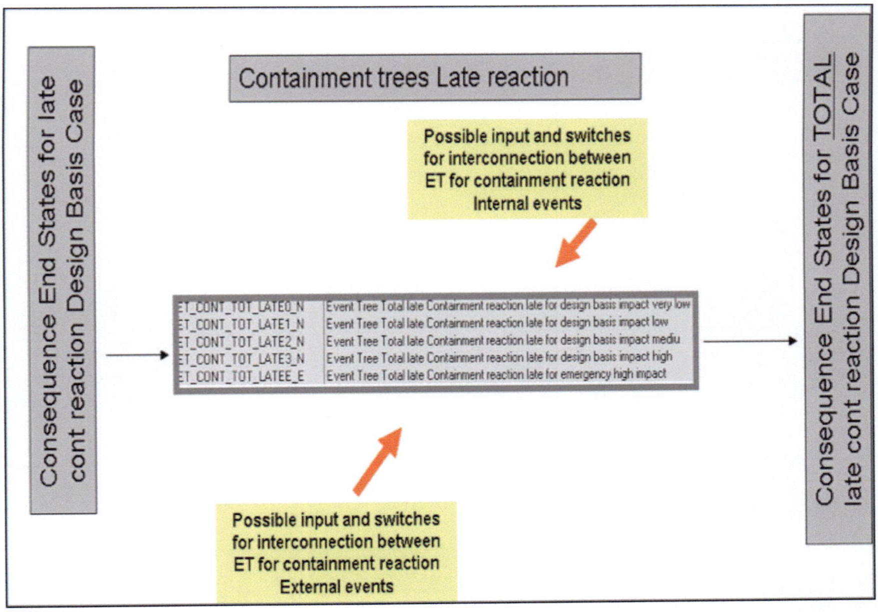

Fig. 2.31 IPSA model—building of the containment ET: 2

  - A good auditable process to establish and manage the assumptions and limitations of the study;
  - Availability and clarity of the input information (from design, operation and/or previous studies);
  - Trained and experienced team in all PSA tasks and/or efficient use of the support/subcontracting teams.

- Experience in using high-performance PSA techniques, as, for instance, management of model magnitude by the use of complex logical equations and conditions.

This example addresses the general managerial approaches in an (I_IPSA_EPSA) study, which has a high impact on all the PSA tasks, but the integration and quantification are the parts of the highest impact.

The managerial aspects of PSA study is related to the need for a clear definition of the objectives. This will identify the target level of quality, on which the depending on which the use of the results is possible):

- If the intended use of PSA is to support risk decisions on the NPP and/or activities related to it (licensing, evaluation of environmental impact, etc.), then a high level of quality and trustfulness of the whole process and model are required. This is achieved by compliance with the existing standards on quality assurance for the PSA tasks;

- The same quality assurance processes are implemented for all the study tasks, as per existing standards [1, 7–15], etc. The study develops a project management approach, with detailed procedures and tasks/responsibilities definition. An example of the need for such a definition is represented in Table 2.8 and Fig. 2.32 [1] for the case when the IPSA tasks have to be correlated and coordinated with the EPSA ones in order to build an (I_IPSA_EPSA) model.

The implementation of a Quality Assurance Manual for the study is also one important condition to have a model and a process, which are auditable. This is important for the post-study activity, which is mentioned in the corresponding paragraph. This includes the existence of auditable and trustful information (from design, operation and other studies) as an input, as mentioned in the first set of problems identified for a PSA study.

Of highest importance is also the assurance of a trained team in all PSA tasks and/or efficient use of the support/subcontracting teams.

Example 2 solution $\mathbf{S}_8^{\mathbf{PR}_8}$: The optimization of the (I_IPSA_EPSA) model is also highly influenced by the ability of the PSA team in using high- performance PSA techniques, as, for instance, the use of complex logical equations and conditions.

SF are used in cases when there is a certain degree of epistemic uncertainty on some decisions points in the PSA study. SF may be used in ET and FT for FE or for deriving IE.

The use of SF for epistemic uncertainties in defining the probability that a certain barrier will be successful was presented in Table 2.2. This type of uncertainty is encountered mostly in case of new designs and/or modifications on which no supporting information is available. The use of such SF is represented in a sample case in Fig. 2.33.

The FE represented in Fig. 2.34 is a case of a BC [1] defined in the ET, which assumes a set of combinations for the switches for internal events in various calls:

- If the call is from an ET on 'loss of DC', then the DC module is switched off in the FT;
- If the call is from an ET on 'loss of AC power', then the AC part of the FT is switched off;
- In both cases, for the control rods action (for which there is lacking information) an SF is included for further consideration if needed in the SUA task.

Example 3 solution $\mathbf{S}_8^{\mathbf{PR}_8}$: The use of combined switches in the FT considering both the support systems and external events parts. The example is presented in two steps:

- The first describing which switches are introduced, as a detailed information to illustrate the general rules for the use of switches, as shown in the examples 2 for solution $\mathbf{S}_7^{\mathbf{PR}_7}$ and 3 for solution $\mathbf{S}_8^{\mathbf{PR}_8}$;
- The second describing details on how the switches impact on the FT model.

Table 2.8 Sample representation of the flow to build I_IPSA_EPSA model

No	Code	Step	Short description	Responsibility
1	IPSA_BEL	Derive the initial list of basic events (BEL)	Derive BEL from the Internal PSA (IPSA)model	IPSA team
2	SEL	Develop Seismic Equipment List (SEL)	Based on IPSA list from step 1—of the BE. The list includes the seismic failures values based on fragility analysis and previous Hazard Analysis (HA) and the Seismic Basic Events (SBE) from IPSA	EPSA team input to IPSA team and agreed
3	IE_MATRIX	Define the interface matrix for each External Initiating Event (Ex-i)	Define the list of the IPSA Initiating Events (IE) and the Function Event (FE) affected due to each EX-i	EPSA team input to IPSA team and agreed
4		Include external event part in the IPSA model		
4.1	Ex-i_IE-IPSA connect	Include Ex-i connections with IPSA IE	Develop Event Trees (ET) for Ex-i and include the logic connectors between Ex-I and the IE of the IPSA as connecting end states in the Ex-i	IPSA team
4.2	ExBE_in IPSA	Include ExBE in the FE	Use logic construction to include ExBE in the IPSA structure by identifying the top closest gate to the internal BE where an ExBE has to be inserted (as logic switches to be activated for each Ex-i case): –including the Boundary Conditions (BC) in the call of each FE in each IE of the IPSA structure. BC are defined as logic switches to be activated for each Ex-i to be run –the support systems that have to be deactivated for a given Ex-i are switched off using logic switches for the parts that have to be decoupled	IPSA team
5	RUN_CASE	Define the case calculation	Define the BC (List of logic switches to be activated) for the case and assign to the calculation case	IPSA team

(continued)

Table 2.8 (continued)

No	Code	Step	Short description	Responsibility
6	RES_REVIEW	Results review	Review results Rank results using probability of Minimal Cut Sets (MCS) and Importances of BE in the MCS Review the main ranked MCS check their meaning Define the main issues for Sensitivity and Uncertainty Analysis (SUA)	IPSA and EPSA teams
7	SUA	Perform S&UA	Perform analysis and define the dominant factors and final ranking of the MCS and contributors	IPSA and EPSA
8	REPORT	Develop report and documentation of the review	Develop report and review quality registrations for the study	IPSA and EPSA

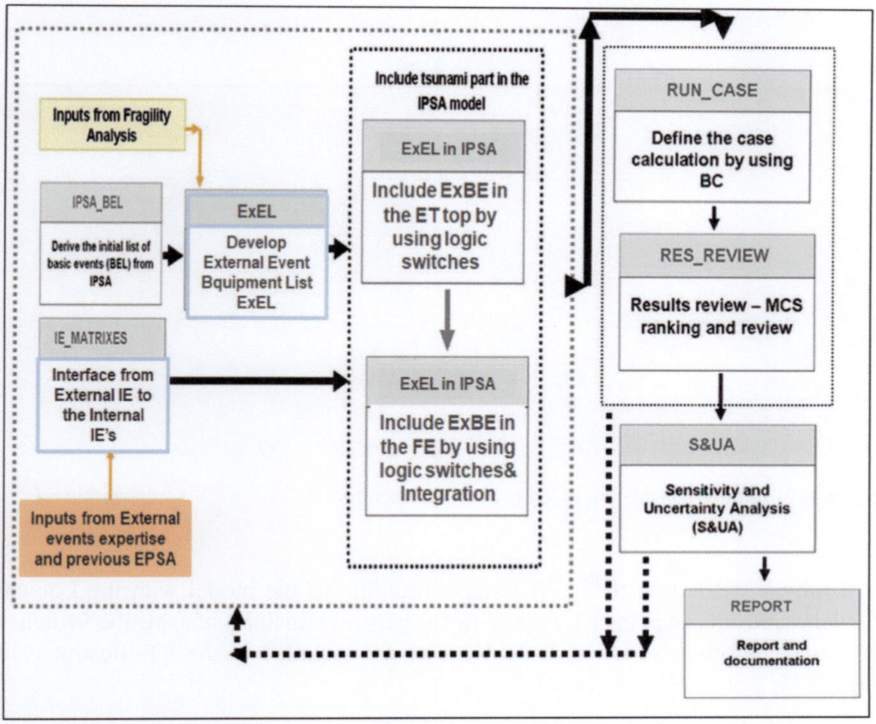

Fig. 2.32 Flow path of inserting external events part into internal events PSA

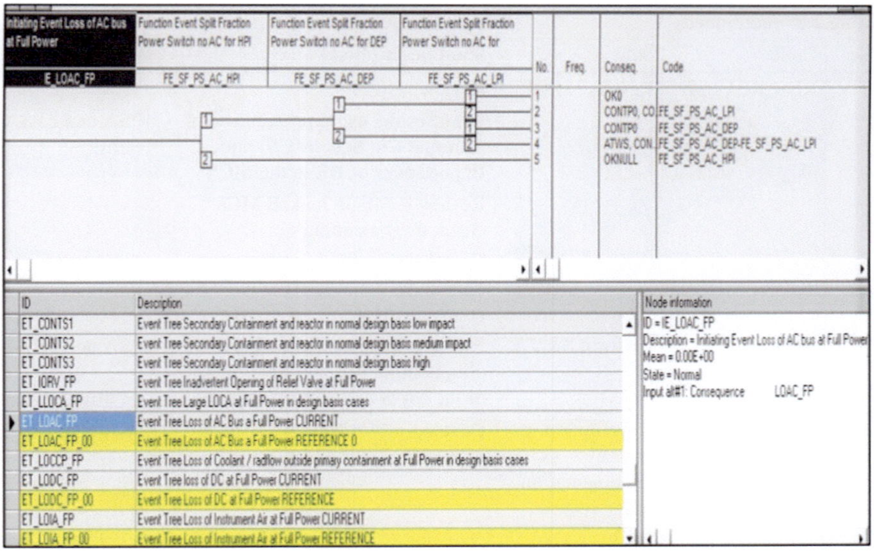

Fig. 2.33 Event tree split fraction use—sample

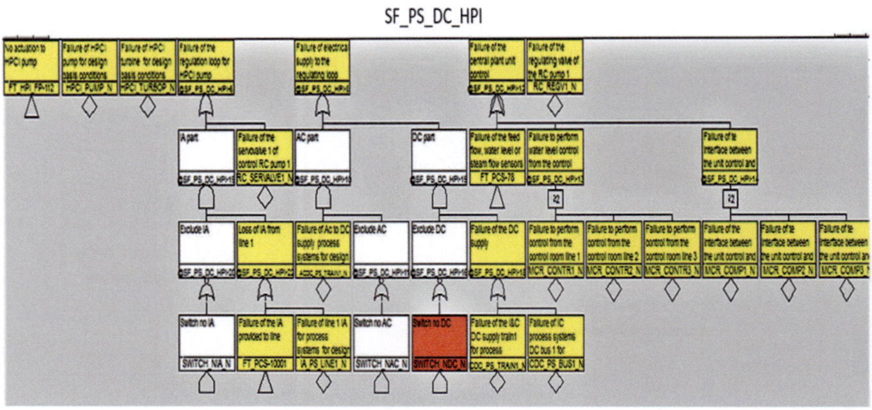

Fig. 2.34 Fault tree considering switches and split fractions

Example 3 solution $\mathbf{S}_8^{\mathbf{PR}_8}$: First step—Building of the model with the Logical Switches introduced in the FT. Following the general principles of using the Switches presented before, a series of logical events are included in the FT illustrated in Figs. 2.35, 2.36, 2.37, 2.38 and 2.39:

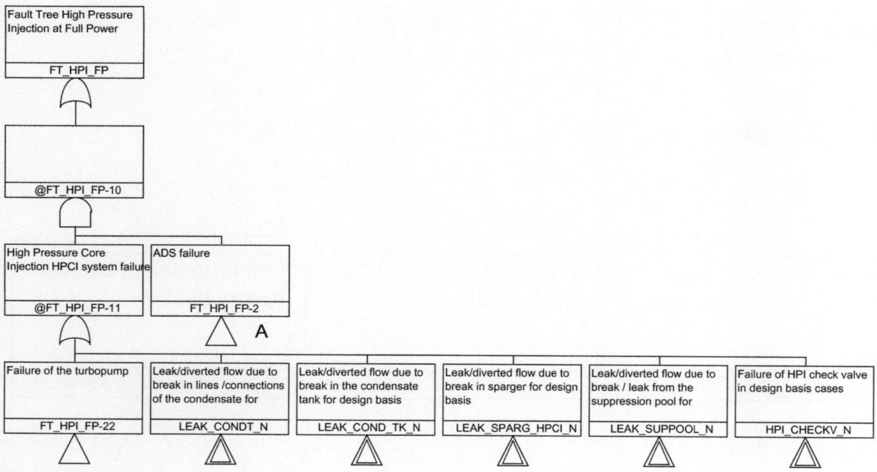

Fig. 2.35 Use of switches in the FT—an example of FT and places were the switches will be included—first level without support systems

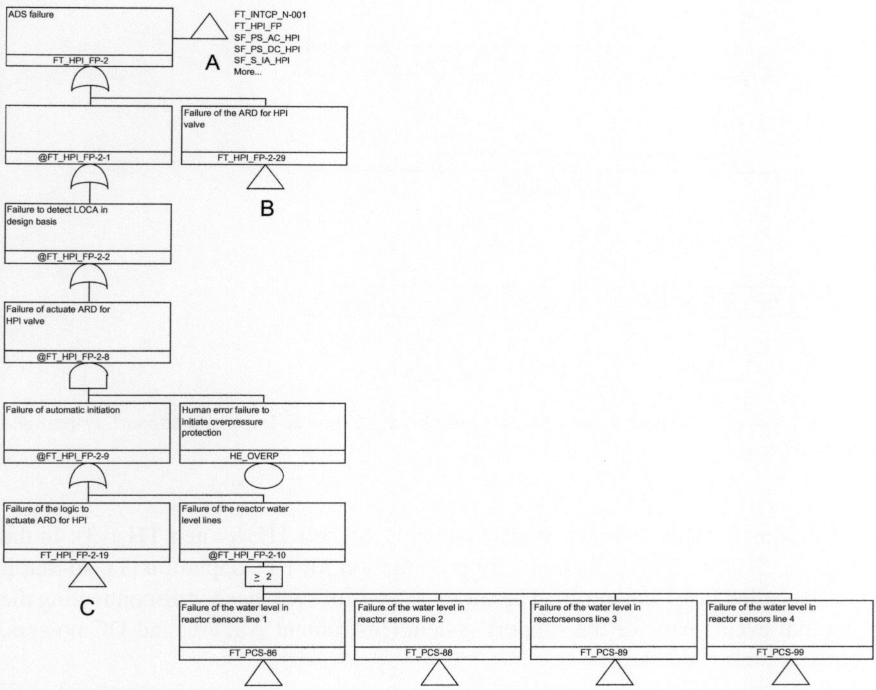

Fig. 2.36 Use of switches in the FT—an example of FT and places were the switches will be included—external level with example of support systems

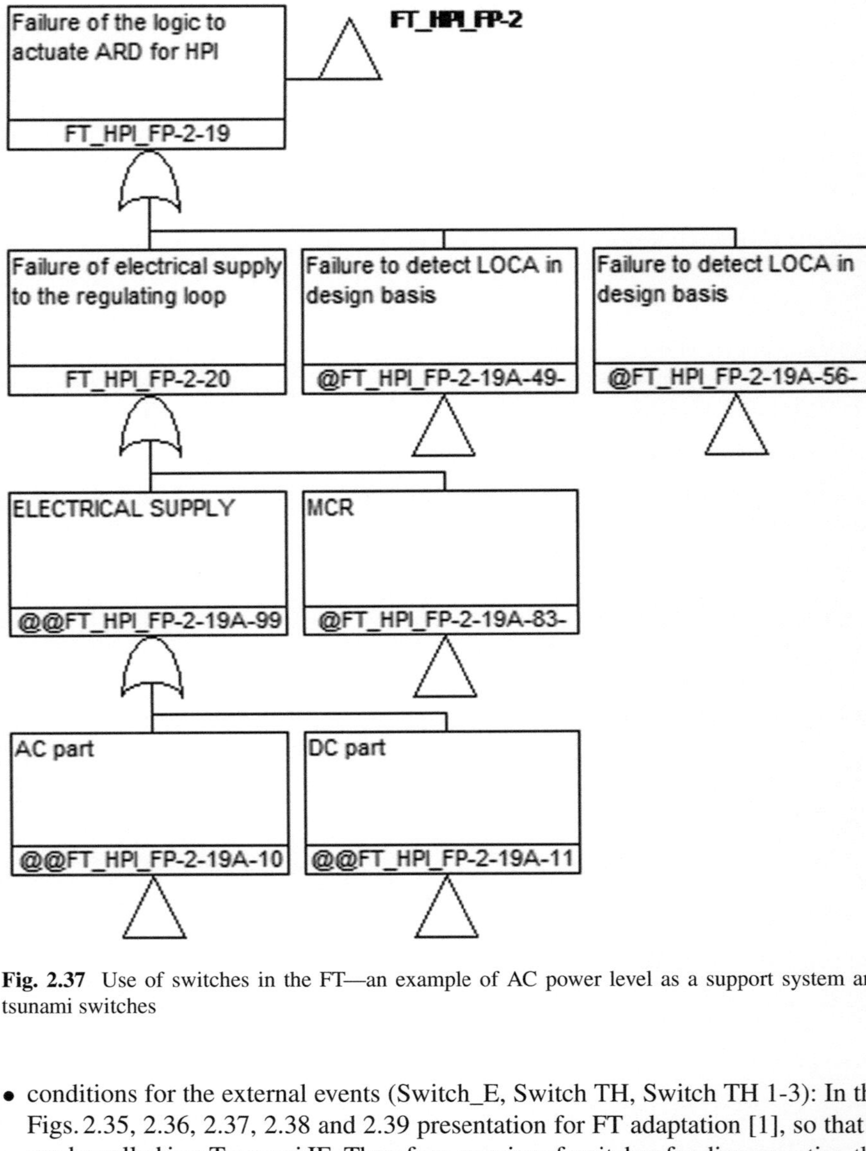

Fig. 2.37 Use of switches in the FT—an example of AC power level as a support system and tsunami switches

- conditions for the external events (Switch_E, Switch TH, Switch TH 1-3): In the Figs. 2.35, 2.36, 2.37, 2.38 and 2.39 presentation for FT adaptation [1], so that it can be called in a Tsunami IE. Therefore, a series of switches for disconnecting the internal events part for the support system (Instrument Air, AC and DC powers, etc.) are to be included;
- introduction of Tsunami-specific BE under the tsunami switches for the level of IE considered (Tsunami is disconnecting various support systems at various levels);
- introduction of the switches for the support systems has to be done at the proper level, as shown in the next figures and this action is actually extremely important in the results of the evaluations. A common systematic error in using switches is that

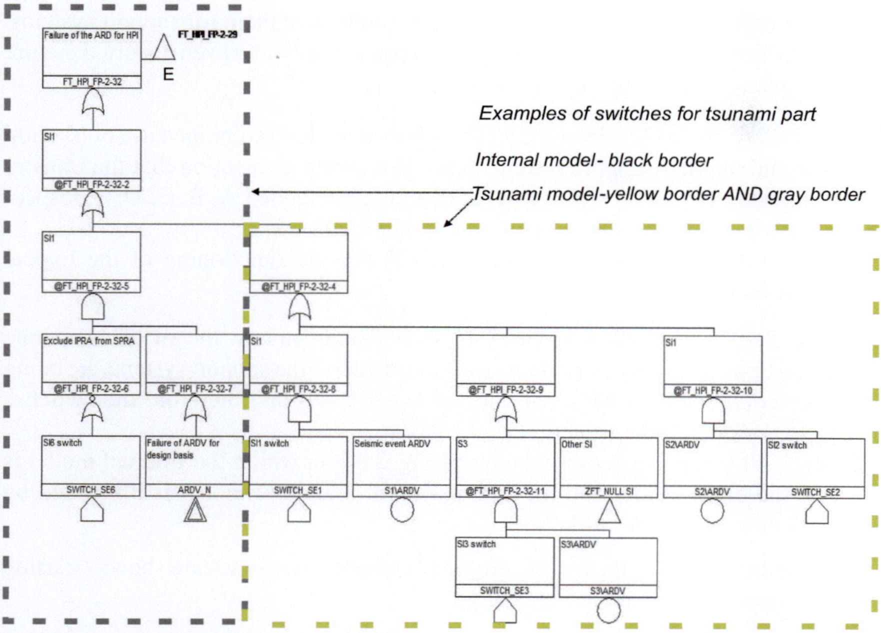

Fig. 2.38 Use of switches in the FT—an example of IA level as a support system and external event switches

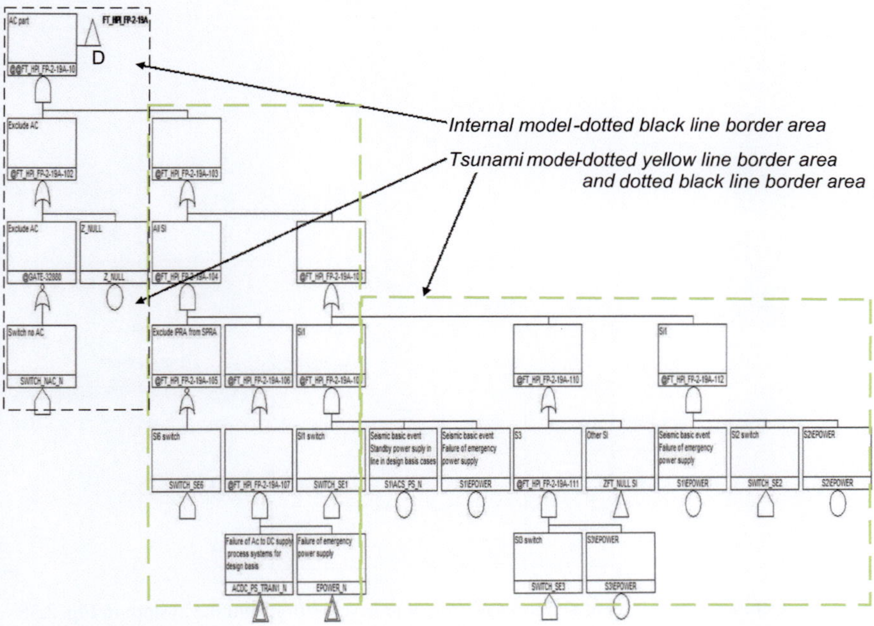

Fig. 2.39 Use of switches in the FT—an example of ACA level as a support system and external event switches

of a poor identification of the level where to implement them for support systems, action which results in very simplified/over conservative and even distorted picture of the plant reaction and contributors.

Figures 2.35, 2.36, 2.37, 2.38 and 2.39 show how to include combined use of tsunami switches and internal support systems ones. It is worth to mention that the transfer gates, shown in Figs. 2.35, 2.36, 2.37, 2.38 and 2.39 labelled A, B, C, D, E, respectively, are top gates that link to one or more other fault trees.

Example 3 solution $\mathbf{S}_8^{\mathbf{PR}_8}$: Second step—Verify the functioning of the logical conditions in the FT.

- Figures 2.40, 2.41, 2.42, 2.43 and 2.44 show details on how the switches operate and illustrate also a very important issue to consider—the support systems are being decoupled/affected by TPSA at various TsE-I levels and therefore the switches have to consider these aspects [1];
- Figure 2.40 is another representation of Fig. 2.35, in which the Internal model is inside the black border and the external event (tsunami) model is illustrated by both yellow and black borders.

For the illustration of the use of combined switches, two cases are shown (starting from the system presented in Fig. 2.35).

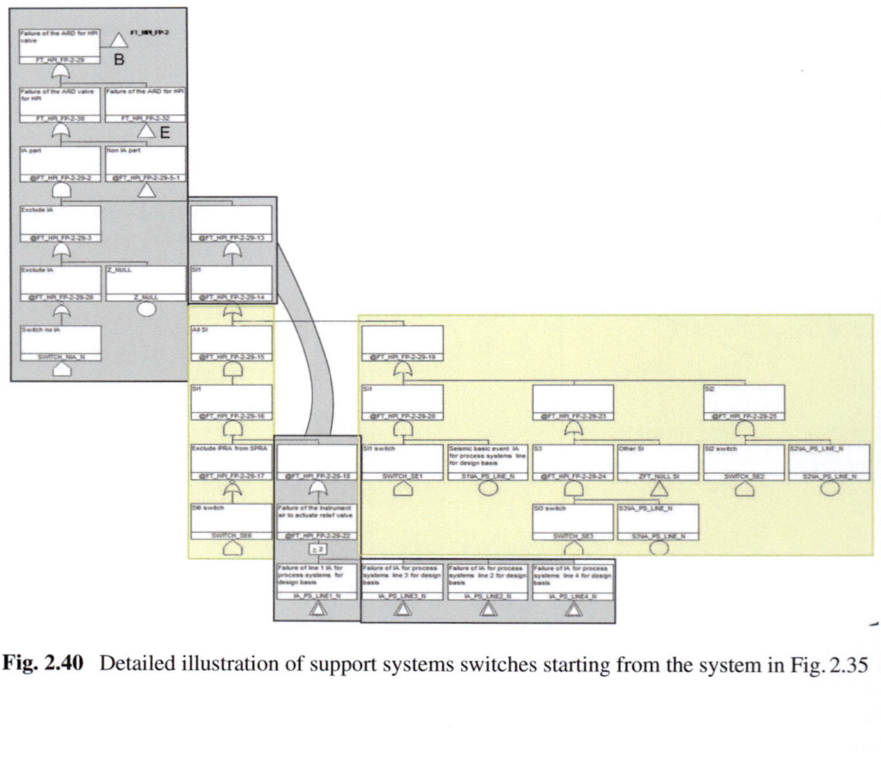

Fig. 2.40 Detailed illustration of support systems switches starting from the system in Fig. 2.35

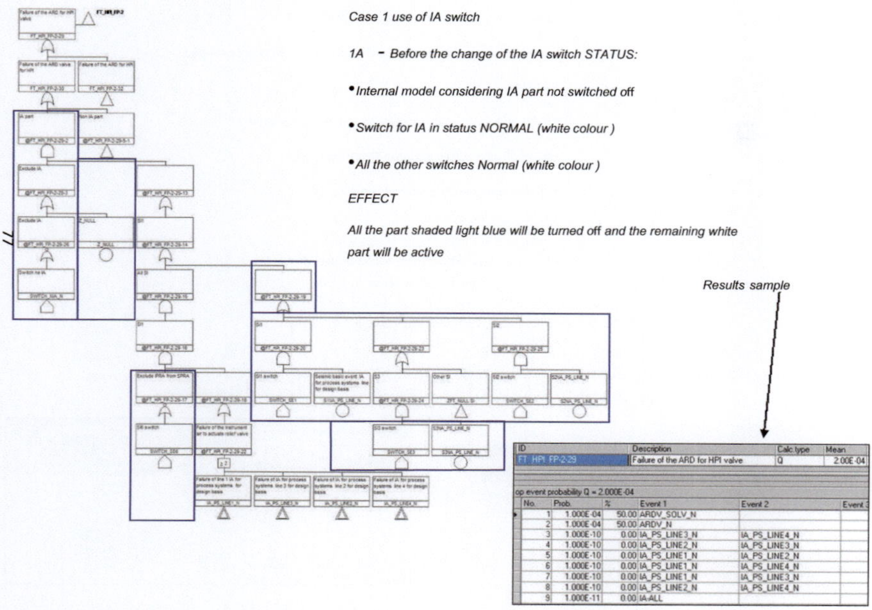

Fig. 2.41 Case 1: the use of the IA switch—impact on sample case from Fig. 2.35. Situation before the use of the IA switch

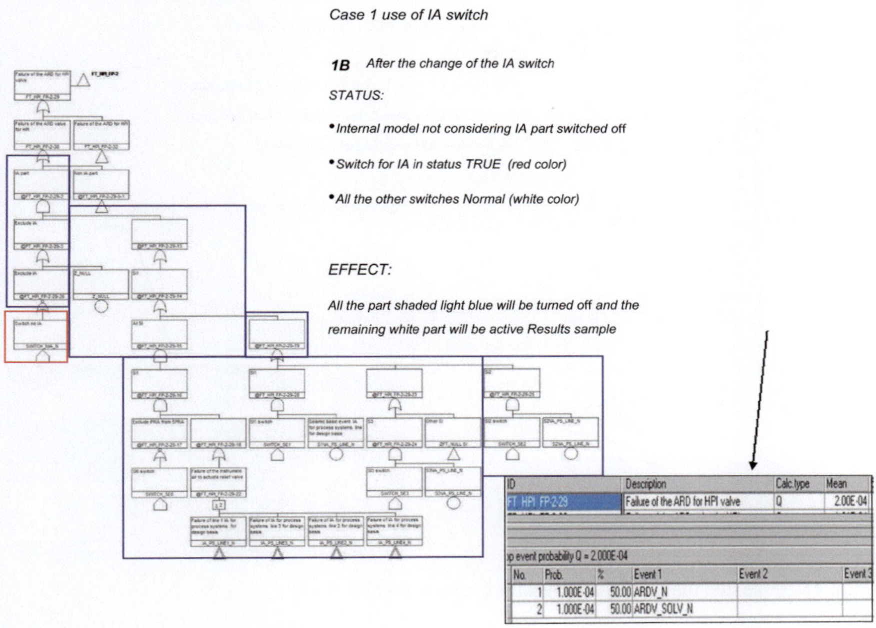

Fig. 2.42 Case 1: the use of the IA switch—impact on sample case from Fig. 2.35. Situation after the activation of the IA switch

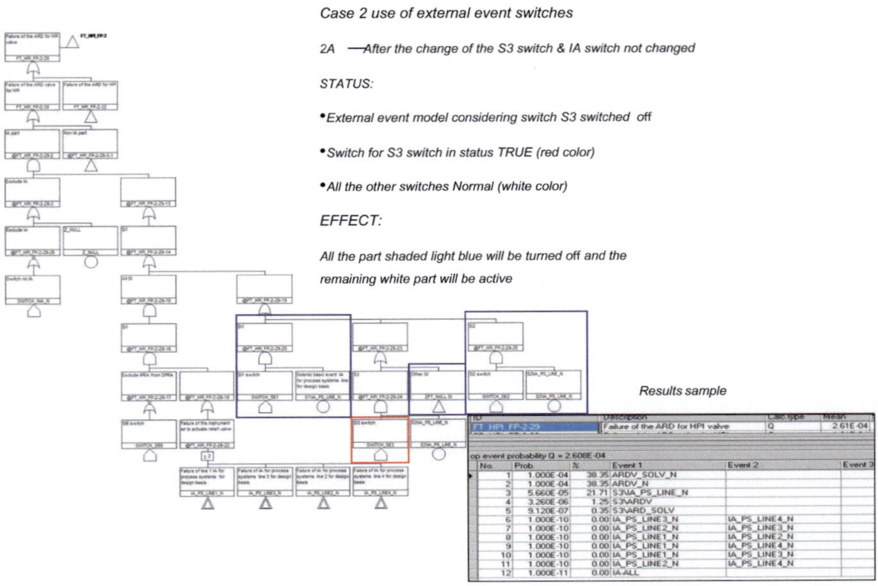

Fig. 2.43 Case 2A: the use of switches for external events and not for IA

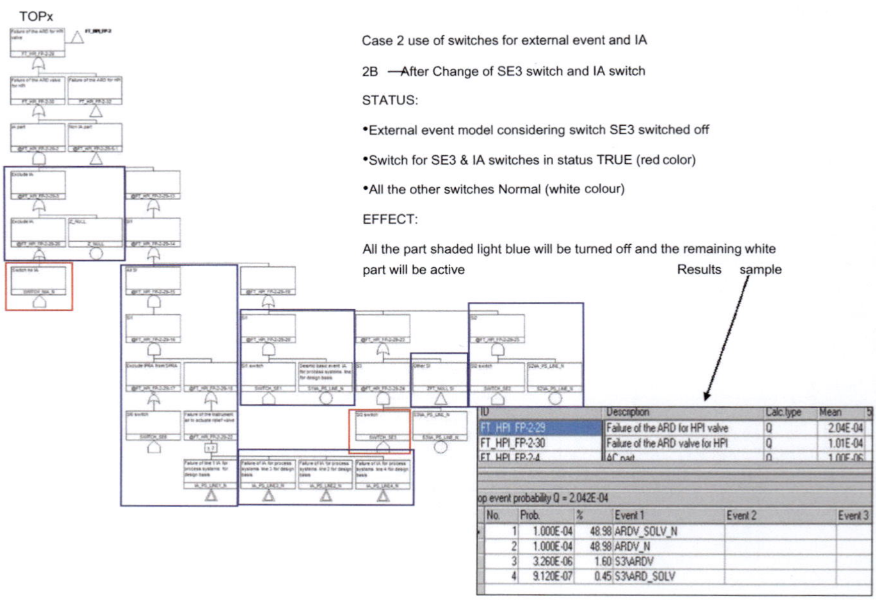

Fig. 2.44 Case 2B: the use of switches for external event and IA

Case 1 illustrating the use of IA switch (Fig. 2.41) with two situations (1A before the activation of the IA Switch and 1B after its activation):

1A—Before the change of the status of IA switch.

- STATUS
  - Internal model considering IA part not switched off;
  - Switch for IA in status NORMAL (white colour);
  - All the other switches NORMAL (white colour).
- EFFECT
  - All the parts inside the blueprint border will be turned off and the remaining white part will be active.

The results for case 1A as for the TOP represented in Fig. 2.41 are shown in the Table 2.9 [1].

1B—After the activation of the IA switch.

- STATUS
  - Internal model considering IA part not switched off;
  - Switch for IA in status TRUE (red colour);
  - All the other switches NORMAL (white colour).
- EFFECT
  - All the parts inside the blueprint border will be turned off and the remaining white part will be active.

The results for the case 1B, after the IA switch activation, are shown in Table 2.10 [1].

Case 2 illustrating the use of external events switches (Fig. 2.43) *with two situations*: (1A before the activation of a combination of internal–external switches and after its activation).

Table 2.9 Case 1A—sample top before the use of the IA switch

TOPx = Failure of ARDV				
1	1E-4	50	ARDV_SOLV_N	
2	1E-4	50	ARDV_N	
3	1E-10	0	IA_PS_LINE3_N	IA_PS_LINE4_N
4	1E-10	0	IA_PS_LINE2_N	IA_PS_LINE3_N
5	1E-10	0	IA_PS_LINE1_N	IA_PS_LINE2_N
6	1E-10	0	IA_PS_LINE1_N	IA_PS_LINE4_N
7	1E-10	0	IA_PS_LINE1_N	IA_PS_LINE43_N
8	1E-10	0	IA_PS_LINE2_N	IA_PS_LINE4_N
9	1E-11	0	IA-ALL	

Table 2.10 Case 1B—sample TOP after IA switch activation

TOPx = Failure of ARDV if Switch IA is activated			
1	1E-4	50	ARDV_SOLV_N
2	1E-4	50	ARDV_N

2A—After the change of EE3 switch & IA switch not changed.

- STATUS
  - External event model considering external event switch EE3 are switched off;
  - Switch for EE3 in status TRUE (red colour);
  - All the other switches NORMAL (white colour).
- EFFECT
  - All the parts inside the blueprint border will be turned off and the remaining white part will be active.

The results for case 1A after the change of EE3 switch & IA switch not changed are shown the Table 2.11 [1].

2B—After the change of EE3 switch & IA switches.

- STATUS
  - External event model considering external event switch EE3 are switched off;
  - Switch for EE3 & IA in status TRUE (red colour);
  - All the other switches NORMAL (white colour).
- EFFECT
  - All the parts inside the blueprint border will be turned off and the remaining white part will be active.

Table 2.11 Case 2A—sample TOP after the change of EE3 switch & IA switch not changed (from Fig. 2.43)

Case 2A TOPx = Failure of ARDV				
1	1E-4	38.35	ARDV_SOLV_N	
2	1E-4	38.35	ARDV_N	
3	5.66E-5	21.71	IA_PS_LINE3_N	IA_PS_LINE4_N
4	3.26E-6	1.25	IA_PS_LINE2_N	IA_PS_LINE3_N
5	9.21E-7	0.35	IA_PS_LINE1_N	IA_PS_LINE2_N
6	1E-10	0	IA_PS_LINE1_N	IA_PS_LINE4_N
7	1E-10	0	IA_PS_LINE1_N	IA_PS_LINE43_N
8	1E-10	0	IA_PS_LINE2_N	IA_PS_LINE4_N
9	1E-11	0	IA-ALL	

2.7 Uncertainty and Sensitivity Analyses

As it was presented in the introduction, PSA can be defined as a complex system (CAS) (Figs. 2.16, 2.17 and 2.18). It was shown in previous works that, during the development of an I_IPSA_EPSA model, an algebra is built and the metrics of it define the risk metrics (CDF, LERF, Risk). This is reflected and in full accordance with the computer codes models of I_IPSA_EPSA.

In the meantime, as Fig. 2.25 and the comments on it mentioned before showed, the model of PSA as a complex system of CAS type involves the approach on the credibility of the results on metrics, which will consider that PSA is a composed of a set of tasks, connected between them, each of them with a certain level of credibility.

For the PSA model, the credibility is considered from the very beginning in the study, as, for instance, in the form of SF (Example 1 of solution $\mathbf{S}_1^{\mathbf{PR}_1}$). The study starts with the process of identification of assumptions and considers continuously this aspect, in order to evaluate the impact of low credibility decision points.

The SUA task of PSA is designed to review the impact of the assumptions on the calculated risk metrics. However, during this process there are some main principles to be considered:

- PSA model is a decision logical construction;
- The elements of the PSA model are probabilistic components.
- For the PSA components the following approach is to be considered: the judgement on the credibility of the results will follow the guidances for deterministic and probabilistic analyses:
  - For the deterministic reasoning: If X is requiring Y to produce the effect W and the two conditions are fulfilled then W will take place;
  - For the probabilistic reasoning: Element X known with uncertainty Ux is the requiring element Y known with uncertainty Uy and they are producing a known effect W with uncertainty Uw;
  - Nevertheless, the OPEX and real operation, as shown in other knowledge management works are guided rather by paraconsistent logic, which is not favouring, but encouraging the refuse of accepting any logic during the review of a failure scenario.

Therefore, in the final analysis, as it will be mentioned in the next paragraph on the use of the PSA results for the decision-making process, in the input to the decision process the SUA play a significant role.

However, SUA is just one of the tools for the review of the credibility of results. The experience of the PSA team and its experts, of the PSA Project Manager, the use of a quality system and internal and external peer reviews are crucial in having a right balance in the identification of the real important factors, that could impact the results and be useful in NPP-related decisions.

Section 4.1 presents in detail the status of research on SUA from the mathematical support point of view.

- The Key Topic for the Sensitivity and Uncertainty Analyses ($\mathbf{KT}_9^{SUA}$) is to evaluate the credibility of results for the further use in the safety-related decisions and applications;
- The problem for the $\mathbf{KT}_9^{SUA}$ ($\mathbf{PR}_9^{SUA}$) is that the definition of the range of variation of results and their credibility is the dominant problem of the SUA task. The difficulty consists of the nature of PSA model and the tools used to solve it, i.e. the combination of logical constructions and probabilistic distribution of the failing components and of the hazards, as well as the impact of real occurrences as processed from the OPEX feedback;
- The solution for the $\mathbf{PR}_9^{\mathbf{KT}_9}$ ($\mathbf{S}_9^{\mathbf{PR}_9}$) is to develop a model of the risk metrics results and a process to evaluate its sensitivity to various parameters, which have an impact on the accuracy and credibility of the PSA study results.

Example 1 solution $\mathbf{S}_9^{\mathbf{PR}_9}$: The I_IPSA_EPSA model depends on many parameters and the SUA task has to consider this important aspect. The implication is that the SUA results provide a range of variation of the results for risk metrics (R in formula 2.14), as represented in Fig. 2.45 [3] and a set of rules on how to evaluate the departure from the reference.

$$Risk\,metric = R = f(\mathbf{x}, a_i) \tag{2.14}$$

where
$\mathbf{x}$—is the algebraic structure defined by the main variables
a_i—are scalars, defining parameters of the risk metrics R.

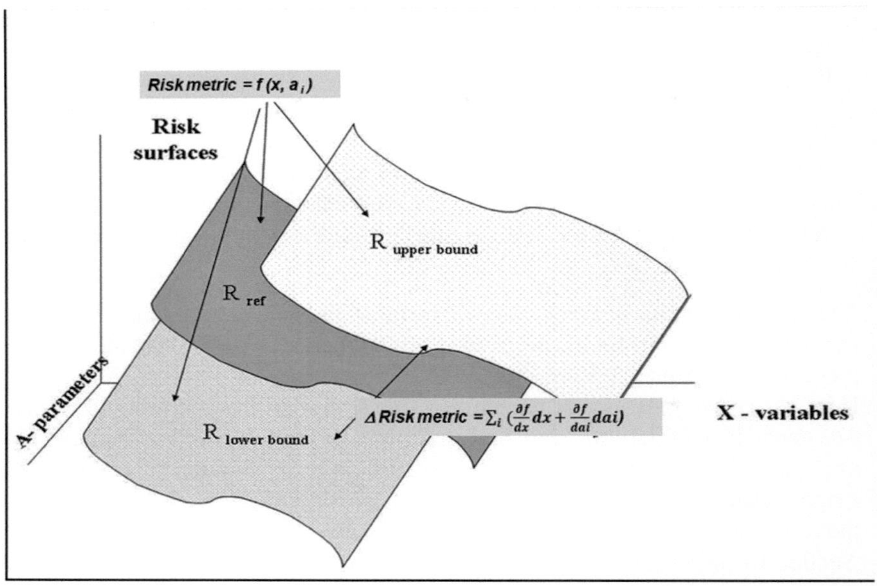

Fig. 2.45 The geometric representation of the risk metrics generated by I_IPSA_EPSA algebra

Table 2.12 [3] represents the manner a set of parameters, which may have an impact on the Risk Metrics can be evaluated, which is leads to two groups of cases, as shown in Example 1 of solution $\mathbf{S}_1^{\mathbf{PR}_1}$:

- one parameter to be variated,
- two or more parameters which are variating.

As a result of applying the rule of one or more parameters variations, for a given level of the I_IPSA_EPSA (levels 1, 2 or 3), the Risk metrics departure from a reference case (Formulas (2.15)–(2.17)) is calculated.

$$\Delta Risk\,metric = dR + dU_R \tag{2.15}$$

$$dR = (\partial R/\partial x)_{a=const}dx + (\partial R/\partial a)_{x=const}da \tag{2.16}$$

$$dU_R = \sqrt{(\partial R/\partial x)^2_{a=const}dx + (\partial R/\partial a)^2_{x=const}da} \tag{2.17}$$

However, due to the fact that the Risk Metrics is assumed to be linear in the logarithmic scale, the simplified evaluations for the departure from the reference $dU + Rcase$ is better represented by formula (2.19) than by formula (2.18).

Table 2.12 Sensitivity analysis cases—sample NPP PSA project

ID	Group I of Sensitivity Analysis—Evaluation of the impact of major assumptions	Parameter 1	...	Parameter N
ID	Group I of Sensitivity Analysis—Evaluation of the impact of major assumptions	Parameter 1	...	Parameter N
0	Base case model	Dummy values optimistic & not correlated between them		Dummy values optimistic & not correlated between them
A_1	Sensitivity case variating parameter 1 by comparison with BASE CASE	Set all values to those figures, configuring the HIGHEST impact of the parameter 1	...	Set all values to those figures, configuring the LOWEST impact of the parameter N
...	...	...	...	...
A_i	Set all values to those figures, configuring the HIGHEST impact of the parameter i and the LOWEST impacts for the rest of parameters			
...	...	...	...	...
A_N	Sensitivity case variating parameter N by comparison with BASE CASE	Set all values to those figures, configuring the LOWEST impact of the PARAMETER 1	...	Set all values to those figures, configuring the HIGHEST impact of the PARAMETER N

$$CRT_1 = dR/dRef \tag{2.18}$$

$$CRT_2 = ln(dR)/ln(dRef) \tag{2.19}$$

Due to the fact, that I_IPSA_EPSA is a CAS, there are various sub-models connected between them. The most important division of those models is based on the general philosophy of PSA, i.e. levels 1, 2 or 3 with their corresponding risk metrics, as described in the introduction. In this case, the evaluation of the uncertainty has specific aspects, correlated also with the fact that the model is developed for and being calculated with a series of specialized computer codes. Both PSA level 2 (L2PSA) and PSA level 3 (L3PSA) use results from the Level 1 PSA (L1PSA). The flow path of the process to develop a full scope levels 1–3 PSA is as shown in Fig. 2.46. Figure 2.46 shows that

- L1 PSA and, respectively, L2 PSA combined with specific inputs for those steps in PSA evaluation, and also different codes combined between them. The calculation of the output of L3 PSA is connected with the output from calculations for L1 PSA and uncertainties at each phase L2 PSA and L3 PSA
- O_i is the output of PSA level i ($i = 1, 2, 3$)
  - O1 is the result for PSA level 1 (CDF) and it is input to PSA level 2. Core Damage States (CDS) are grouped as it is presented in the next paragraph on the specifics on PSA level 2 and they are input to PSA level 2;
  - The result of PSA level 2 (O2) is characterized by LERF (and in some new designs RC- Release categories (see) O2 are input to PSA level 3, for which the result is O3 (Risk).

For all this flow path an overall level of credibility is accompanying the risk metrics outputs (CDF, LERF/RC, Risk), as presented in Fig. 2.46 [16, 17].

As shown in Fig. 2.46, the uncertainty calculations for the case of using different codes in PSA may be computed as

$$O_1 + \Delta U_3 \equiv O1 + \Delta U_{TOT} = O1 + f(\Delta U_1, \Delta U_2, \Delta U_3) \tag{2.20}$$

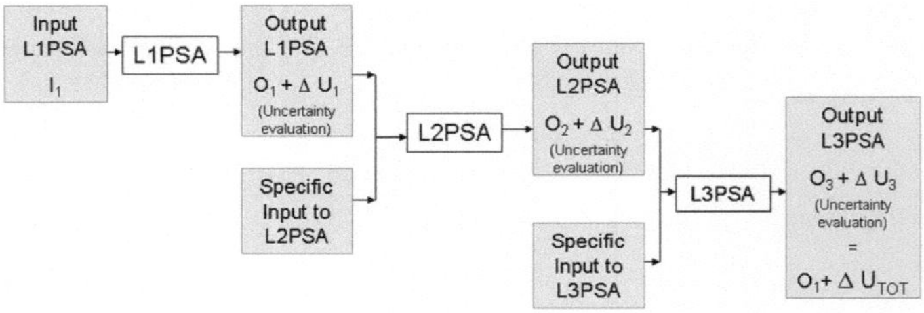

Fig. 2.46 PSA flow path from the credibility/uncertainty point of view

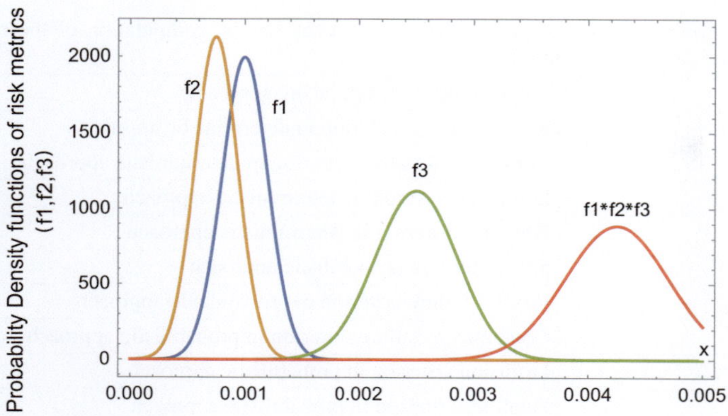

Fig. 2.47 Representation of the convolution integral for total distribution of the risk Metrics for I_IPSA_EPSA levels 1–3 integrated

It is worth to mention that the risk metrics curves are fundamental of probabilistic nature and their combination needs to be evaluated after calculating convolution integral of the resultant final risk metrics curve. The process is represented for levels 1–3 PSA in formula (2.21) and Fig. 2.47.

$$f = (f_1 \circledast f_2 \circledast f_3)(x) \tag{2.21}$$

where f_1, f_2, f_3 are the densities probability of risk metrics for CDF, LERF and RISK, respectively.

Example 2 solution $\mathbf{S}_9^{PR-9}$. In the previous example, a generic situation of I_IPSA_EPSA risk metrics SUA case was presented. However, PSA is performed not only for the evaluation of the risk metrics, but also for various applications. One of those applications consists of the definition of radii of the protection zones around NPP.

In such case, the uncertainties follow the path of calculations from formulas (2.22) to (2.24).

For the sake of underlying the computational aspects of the radii in a deterministic and probabilistic approaches, coded by indexes 'd' and 'p', respectively, a set of formulas can be derived as presented below, for the variables introduced in Table 2.13:

$$Radii_d = S_d \cdot R_d \cdot C_d \cdot Diff_d \cdot D_d \pm \Delta U_d \tag{2.22}$$

$$Radii_p = S_p \cdot R_p \cdot C_p \cdot Diff_p \cdot D_p \tag{2.23}$$

Further, we denote f_1 the density function for the probabilistic criteria for S_p, f_2 t the density function for the probabilistic criteria for R_p, f_3 the density function for

Table 2.13 Deterministic and probabilistic approaches for the computation of the radius/radii size(s) around a nuclear power plant

Sd	Source term in deterministic approach
Rd	Reactor failure criterion in deterministic approach
Cd	Containment failure criterion in deterministic approach
Diffd	Diffusion criterion in deterministic approach
Dd	Fatalities criterion in deterministic approach
Sp	Source term in probabilistic approach
Rp	Reactor failure criterion in probabilistic approach
Cp	Containment failure criterion in probabilistic approach
Diffp	Diffusion criterion in probabilistic approach
Dp	Fatalities criterion in probabilistic approach
ΔU_d	Uncertainties in deterministic approach
f_1, f_2, f_3, f_4, f_5	Distribution functions for the probabilistic criteria
F	Convolution of functions f_1 to f_5

the probabilistic criteria for C_p, f_4 the density function for the probabilistic criteria for $Diff_p$, f_5 the density function for the probabilistic criteria for D_p, and $\circledast$ the convolution operator.

Then, the convolution integral of the probabilistic criteria leads to

$$F = \int_{\mathrm{R}} (f_1 \circledast f_2 \circledast f_3 \circledast f_4 \circledast f_5)(p)dp \tag{2.24}$$

In PSA studies, the best recommended approach on making the difference and defining the threshold between sensitivity and uncertainty analyses is (as per) the following: if the sensitivity analysis shows that the level of impact on the risk metrics is less than one order of magnitude, then detailed uncertainty analyses, as provided by mathematical statistical support is applied. For more details, the readers are referred to Sect. 4.1.

Example 3 solution $\mathbf{S}_9^{\mathbf{PR}_9}$. This example shows the importance of using diverse tools for SUA of the PSA results. In the case when the final results are interesting not only from the point of view of generic Risk metrics values, but also to identify weak points of the NPP, then diverse methods for SUA might be used.

As shown in the Sect. 4.1, the mathematical evaluation of the ranking of elements in a PSA result (is to use criteria as, for instance, 'Importance Measures').

Importances are defined in various ways, but the common feature is that they try to consider the impact of a component that appears in many MCS of the risk metric.

The results of the rankings need to consider not only the probabilities of the MCS, but also the contributions and importance for the contributing components failures.

Table 2.14 Sample case of sequences for SUA ranking: method A

*Seq*1	$IE1 \star B1 \star T1$	$IE2 \star B2 \star T2$	
*Seq*2	$IE2 \star B1 \star T3 \star B2$	$IE2 \star B4 \star T1 \star B2$	$IE2 \star B3 \star T2$
*Seq*3	$IE3 \star B2 \star T3 \star B1$	$IE3 \star T2 \star B3$	

Table 2.15 Sample case of sequences for SUA ranking: method B

*Seq*1	$Seq_1 = IE1 \star B1 \star T1 \dotplus IE2 \star B2 \star T2$
*Seq*2	$Seq_2 = IE2 \star B1 \star T3 \star B2 \dotplus IE2 \star B4 \star T1 \star B2 \dotplus IE2 \star B3 \star T2$
*Seq*3	$Seq_3 = IE3 \star B2 \star T3 \star B1 \dotplus IE3 \star T2 \star B3$

A detailed example of such results is in Table 2.14. This use of SUA task is performed as part of the PSA risk metrics and overall results post-processing. In order to rank the impact of various contributors, there are two possible approaches:

- One that considers the value of the probability of the sequence and uses expert opinions to evaluate possible other cases of importance with low probabilities (Method A);
- Another one that using a combined set of criteria for ranking (not only the probability). Combination of criteria and ranking can be done using existing methods in mathematics, as, for instance, the multi-criteria decision analyses (Method B).

The result of quantification is a list of sequences and their components and the probability, for which methods of groups A are used. The use of method A leads to a certain ranking and the use of expert opinion may not be always traceable and auditable. Therefore, a possible improvement could be brought by methods of group B especially in the case of PSA-specific case (for instance, TsPSA) when the peer review, experience and practice are yet in the beginning.

One possible approach illustrated in Table 2.14 may use not only the probabilities of sequences, but also the probabilities and importance of the constituent events, so that the ranking of the sequence is more refined and the chances to lose significant contributors leading to low-probability sequences is decreased and can be iterated and audited easier.

An illustration for the implementation of those methods is presented in Tables 2.14 and 2.15. A sample in Table 2.14 of sequences is assumed. In those sequences, the IE are initiating events, B are random failures (non tsunami related), T are random failures (tsunami related). In Table 2.15, the ranking considers only the frequencies of the sequences [1, 3, 18].

In Table 2.14, the following colour ranking code has been be adopted:

- Red → High impact
- Yellow → Medium impact
- Green → Low impact

However, if the Method B is applied, then the ranking will change, as shown in Table 2.15, due to the contribution of high importance components even in lower frequency sequences:

As a result of those rankings further iterations in the model and SUA priorities for more evaluations and refinements can lead to the following conclusions:

- For the feedback based on methods A, the dominant issue is given by
  - Seq 1 as a whole (with the combinations leading to it) and
  - the elements defining it: IE1, B1, T1, B2, T2, T3 (having all the same weight).
- For the feedback based on methods B, the dominant issue is given by
  - Seq 2 as a whole (with the combinations leading to it) and
  - the elements defining it.
- B2, B4, T1, T2: Group I of importance
- B3, IE2, T3: Group II of importance
- B1: Group III of importance (lowest)

The results and the rankings in each of the cases above lead to different actions. So for methods B, the further SUA will be focused on other priorities than for methods A, i.e. by giving priority to the whole Seq2 and B2, B4, T1, T2.

Another issue to be mentioned is that the comparative evaluation of the contributors for IPSA and of the dominant ones in the case of TPSA may show the fact that there are changes in the ranking of dominant elements and systems.

Similar evaluations considering other hazards, as, for instance, seismic considerations for TPSA, or multiunit aspects are expected to identify, amongst other things, different contributors based on the specific hazard and/or contributors under consideration. The results are typically presented as mean values considering the uncertainty bands and the impact evaluation of the main parametric values of the risk metric (CDF).

Figure 2.48 shows more details on the evaluation of the risk metrics and sequences by using two SUA methods [1].

	INITIAL CD RANK BY SEQUENCE PROBABILITY	Codes of basic events in MCS of the sequence	RANKING OF ELEMENTS OF SEQUENCES AND SEQUENCE				Description of basic events in MCS of the sequence
			ELEMENT RANK BY PROB	ELEMENT RANK BY IMP F-V	ELEMENT RANK BY COMBINATION	FINAL SEQUENCE RANK	
1		T\OFFPOWER					Failure of Offsite Power due to tsunami event
		T\EPS					Failure of Emergency power supply (Diesels) due to tsunami event
		T\EPS_ALL					CCF forf Emergency power supply (Diesels) due to tsunami event
		T\EPOWER					Failure of DC emergency power supply due to tsunami event
		T\COND_TK_LK					Loss of flow due to leaks in condenser tank after tsunami event
		T\COND_ISO					Failure of condenser isolation system after tsunami event
		T\HE_REPOWER					Human error to recover lost power after tsunami
2		T\OFFPOWER					Failure of Offsite Power due to tsunami event
		T\EPOWER					Failure of DC emergency power supply due to tsunami event
		S3\EPOWER_ALL					CCF for DC emergency power supply due toi tsunami event
		T\COND_TK_LK					Loss of flow due to leaks in condenser tank after tsunami event
		T\COND_ISO					Failure of condenser isolation system after tsunami event
		T\HE_REC_F					Human error to restore flow after tsunami event
3		T\AC_POWER					Failure of AC POWER due to tsunami event
		T\EPOWER					Failure of DC emergency power supply due to tsunami event
		T\EPS					Failure of Emergency power supply (Diesels) due to tsunami event
		T\EPS_ALL					CCF forf Emergency power supply (Diesels) due to tsunami event
		T\SEECOOL					Seacooling tsunami event
		T\COND_TK_LK					Loss of flow due to leaks in condenser tank after tsunami event
		T\COND_ISO					Failure of condenser isolation system after tsunami event
		T\HE_AC_REC					Human error to recover AC after tsunami
4		T\OFFPOWER					Failure of Offsite Power due to tsunami event
		T\EPOWER					Failure of DC emergency power supply due to tsunami event
		T\COND_TK_LK					Loss of flow due to leaks in condenser tank after tsunami event
		T\ESRV					Failure of emergency relief valve to reclose due to tsunami event
		T\PILON					CCF for DC emergency power supply due toi tsunami event
		T\HE_REPOWER					Human error to recover lost power after tsunami
5		T\OFFPOWER					Failure of Offsite Power due to tsunami event
		T\EPS					Failure of Emergency power supply (Diesels) due to tsunami event
		T\EPS_ALL					CCF forf Emergency power supply (Diesels) due to tsunami event
		T\COND_TK_LK					Loss of flow due to leaks in condenser tank after tsunami event
		T\COND_ISO					Failure of condenser isolation system after tsunami event
		T\HE_REPOWER					Human error to recover lost power after tsunami

Fig. 2.48 Sample set of results of dominant cases for a TPSA

References

1. Serbanescu D (2016) A PSA practitioner and safety decision making person view on some issues related to multiple unit PSA analyses. Kick off meeting of the multiunit PSA project work area 3. In: Vienna IAEA. https://doi.org/10.13140/rg.2.2.32906.06082
2. Graan HV, Serbanescu D, Eloff L, Combrink Y (2005) Some lessons learnt from the use of PRA during the design phase. Int J Crit Infrastruct 1(2–3):287–292
3. Serbanescu D (2015) Selected topics in risk analyses for some energy systems. LAP LAMBERT Academic Publishing
4. RiskSpectrum (2019) RiskSpectrum Doc. http://www.riskspectrum.com/en/risk/Meny_2/RiskSpectrum_DOC/RiskSpectrumDocslide-show
5. van Graan H, Serbanescu D, Combrink Y, Coman O (2004) Seismic initiating event analysis for a PBMR plant. American Nuclear Society - ANS, United States. http://inis.iaea.org/search/search.aspx?orig_q=RN:40038040
6. Serbanescu D (2017) On some aspects of the multiunit probabilistic safety analyses models. In: 2017 international conference on energy and environment (CIEM), pp 293–297. https://doi.org/10.1109/CIEM.2017.8120842
7. PRA procedures guide: a guide to the performance of probabilistic risk assessments for nuclear power plants: chapters 9–13 and appendices A-G (NUREG/CR-2300, vol 2). The American Nuclear Society, LaGrange Park, IL 60525 (1983)
8. NUREG - 1150: severe accident risks: an assessment for five U.S. nuclear power plants. US Nuclear Regulatory Commission, Washington, DC (1990)
9. Report NUREG/CR-6172: reviewing PSA based analyses to modify technical specifications at nuclear power plants. US Nuclear Regulatory Commission, USNRC Washington, DC (1995)
10. Regulatory guide 1.175: an approach for plant specific, risk-informed decision-making: in service testing. US Nuclear Regulatory Commission, USNRC Washington, DC (1998)
11. Regulatory guide 1.178: an approach for plant-specific risk-informed decision-making: in service inspection of piping. US Nuclear Regulatory Commission, USNRC Washington, DC (1998)
12. Report NUREG/CR-6141: handbook of methods for risk-based analyses of technical specifications. US Nuclear Regulatory Commission, USNRC Washington, DC (1998)
13. Standard ANSI/ANS-58.21-2007: external-events PRA methodology. American Society of Mechanical Engineers/American Nuclear Society, ASME/ANS, New York (2007)
14. RA-S-2008: standard for level 1/large early release frequency probabilistic risk assessment for nuclear power plant applications. American Society of Mechanical Engineers/American Nuclear Society, ASME, New York (2008)
15. A guide to nuclear regulation in the UK (updated). US Nuclear Regulatory Commission, USNRC Washington, DC (2016)
16. Some specifics of the use of probabilistic risk analyses as a support to the evaluation of safety margins and the interface with the deterministic based decisions. In: Proceedings of the technical meeting on Effective combination of deterministic and probabilistic safety analysis in plant safety management, Paper 29, IAEA (2006). https://doi.org/10.13140/RG.2.1.2794.8647
17. Kubanyi J, Lavin RB, Serbanescu D, Toth B, Wilkening H (2008) Risk informed support of decision making in nuclear power plant emergency zoning, generic framework towards harmonising NPP emergency planning practices. DG JRC Institute for Energy
18. Safety Reports Series (2018) Consideration of external hazards in probabilistic safety assessment for single unit and multi-unit nuclear power plants, No. 92. International Atomic Energy Agency, Vienna. https://www.iaea.org/publications

Chapter 3
Special Topics in Probabilistic Safety Assessments Levels 2, 3 and PSA Applications

Abstract The special topics presented in this chapter are related to the in Probabilistic Safety Assessments (PSA) level 2 (considering failure of the reactor followed by the failure of the containment) PSA level 3 (considering that there will be release to the environment), which are evaluating the risk impact on the NPP site for the workers, and, respectively, for the environment and population. Starting with level 2 PSA the increasing degree of uncertainty in assumptions makes the tasks related to the post-processing and interpretation of results of high interest. From this perspective, some aspects are presented in detail, as, for instance, the interface between the PSA assumptions and models and the general safety paradigms adopted by the international community at a certain moment in time, the use of PSA results for various applications aimed at supporting the improvements in the safety level at NPP and the use of PSA results for the decision-making process on safety aspects. The feedback to the PSA inputs is considered also important, as well as some aspects related to the research activities supporting PSA methodology.

PSA level 2 has some specific differences by comparison with the level 1, which need to be considered. The most important set of such differences is related to the type of challenges for which the model is performed.

- PSA Level 1 is describing the plant reaction on challenges, which are defined by the Design Basis Accidents. The scenarios on how the Core Damage (CD) could appear and progress to the point of starting to release radioactivity, they assume the reaction of plant barriers designed as Special Safety Systems (SSY) and their Support Systems (SupSy), which are largely based on well-proven codes, experiments and OPEX. On the other side, Level 2 is describing a set of accidents, beyond the design basis, (Beyond Design Basis Accidents—BDBA).
- Historically PSA level 1 started its development by the time the concept of DBA and Defence in Depth got large recognition, after the TMI accident. This was a

D. Serbanescu and A. P. Ulmeanu, *Selected Topics in Probabilistic Safety Assessment*, Topics in Safety, Risk, Reliability and Quality 38,
https://doi.org/10.1007/978-3-030-40548-9_3

major safety paradigm change from intrinsic safety of early nuclear accidents to the layers of defence and barriers for a set of accidents assumed by design. On the other side, the PSA level 2 development started after Chernobyl and became of high importance after Fukushima: a new safety paradigm of protecting the NPP to BDBA and extending DBA was started and it is going on. The specifics of the ET in level 2 were also shown in Example 1 solution $\mathbf{S}_4^{\mathbf{PR}_4}$.

- Both the NPP behaviour and the operator model after a BDBA are subject to intensive research activities. They are based on codes and theories under review and partially confirmed. BDBA research is yet to answer questions on the severe accident progression in water reactors with the release of hydrogen, interaction of core melt with the concrete, operator models in such extreme conditions and many others. In the meantime, PSA level 2 are developed based on the best recognized for specific type of NPP's codes and models, acknowledging the fact that in level 2 the epistemic uncertainties are extremely high and special tools and margins are to be assumed in order to have a conservative approach if such results are to be used.

It seems that there are many severe challenges, generating special topics for PSA level 2. Some of those are presented in the next examples.

- *The Key Topic for the Level 2 Modelling* ($\mathbf{KT}_{10}^{GR2}$) is for most of the NPP, represented in Fig. 1.1 and having a containment, the description of the containment reaction to the challenges created by the evolution of the accidents beyond the DBA status.
- *Problem for the* $\mathbf{KT}_{10}^{GR2}$ ($\mathbf{PR}_{10}^{\mathbf{KT}_{10}}$) *is the description of the containment reaction to prevent the release of radioactivity to the environment in case of a severe accident progression of BDBA type.*
  The aspects that require special evaluations based on specialized computer codes, research activities for the severe accidents phenomena, which are related to the
  - combination in a much higher degree than for the PSA level 1 model of the deterministic approaches (code calculations, experiments plant operator model in severe accidents scenarios, etc.);
  - all in the area of issues under development and review and the probabilistic approaches.
- Solution for the $\mathbf{PR}_{10}^{\mathbf{KT}_{10}}$ ($\mathbf{S}_{10}^{\mathbf{PR}_{10}}$) is to adopt a series of NPP containment scenarios based on existing best-known and peer-reviewed calculations and researches. The assumptions introduced in the Containment Event Tree (CET) are based on existing results for the Engineered Safety features assumed to cope with the BDBA in the given plant. Even if the ESF are a matter of development, especially after Fukushima, their review and development introduced new concepts and they are used in the PSA model in general and CET in particular.

Example 1 solution $\mathbf{S}_{10}^{\mathbf{PR}_{10}}$: Preparing the basis for modelling NPP containment reaction to the challenges after CDS (modelled as ES in the PSA level 1) took place.

End States	Seq no	IE	Function Events No success = FE perform the function					Sequence Logic Equation
		IE_CET IE	FE_CET_1 FE_1	FE_CET_2 FE_2	FE_CET_3 FE_3	FE_CET_4 FE_4	FE_CET_5 FE_5	
		IE Level 2	Integrity	No CD	Control Hydrogen	Integrity temp & pres	Vent & makeup	
OK	SEQ 1	Yes	No	No	No	No		$IE^* nonFE_1 * nonFE_2 * non\ FE_3 * non\ FE_4$
	SEQ2	Yes	No	No	No	Yes	No	$IE^* nonFE_1 * nonFE_2 * non\ FE_3 * FE_4 * non\ FE_5$
	SEQ5	Yes	No	Yes	No	No		$IE^* nonFE_1 * FE_2 * non\ FE_3 * non\ FE_4$
	SEQ6	Yes	No	Yes	No	Yes	No	$IE^* nonFE_1 * FE_2 * non\ FE_3 * FE_4 * non\ FE_5$
RC0	SEQ9	Yes	Yes					$IE^* FE_1$
RC1	SEQ7	Yes	No	Yes	No	Yes	Yes	$IE^* nonFE_1 * FE_2 * non\ FE_3 * FE_4 * FE_5$
	SEQ8	Yes	No	Yes	Yes			$IE^* nonFE_1 * FE_2 * FE_3$
RC2	SEQ4	Yes	No	No	No			$IE^* nonFE_1 * nonFE_2 * nonFE_3$
RC3	SEQ3	Yes	No	No	No	Yes	Yes	$IE^* nonFE_1 * nonFE_2 * non\ FE_3 * FE_4 * FE_5$

Fig. 3.1 Logical expressions for RCs

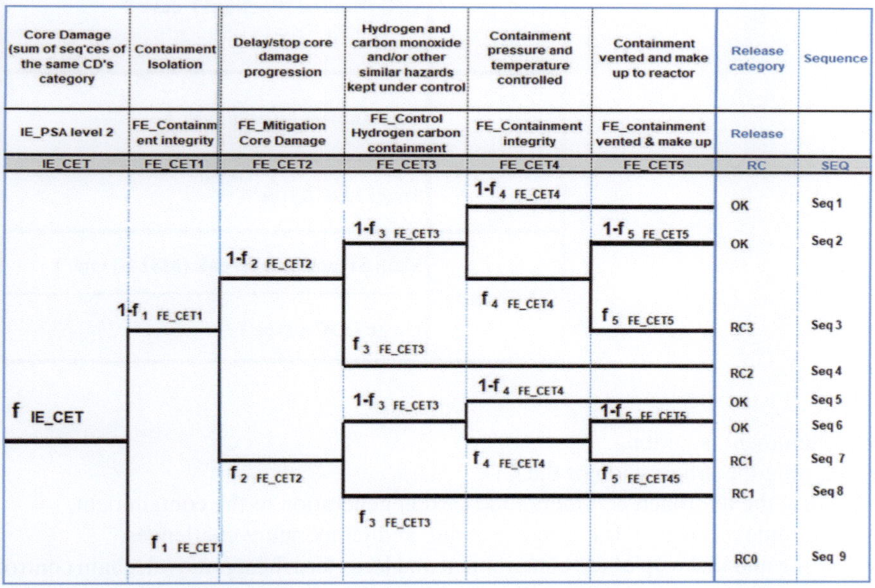

Fig. 3.2 Sample of a typical Containment Event Tree (CET) for a case when PSA level 1 makes sense and has results of risk metrics (CDF)

The input into the PSA level 2 CET is from the PSA level 1, for which all the sequences leading to a certain level of CD are grouped, in a new list of IE, the IE for the PSA level 2. Figures 3.1 and 3.2 illustrated this process of input to the CET.

The FE defined for the CET are based on computer codes calculation on the reaction. These codes are modelled using two major generic approaches (in a phenomenological and/or mechanistical approach) the severe phenomena taking place after CD started. The phenomena are related to the existing means (called ESF) to assure

Fig. 3.3 CD States sample case of risk metrics results after PSA level 1

CD Group 1	Containment by-pass type 1
	Containment by-pass type 2
	Small LOCA type 1
	Small LOCA type 2
	Small LOCA type 3
	SBO (Station Blackout) case 1
	SBO (Station Blackout) case 2
	SBO (Station Blackout) case 3
CD Group 2	Containment by-pass type 3
	Small LOCA type 4
CD Group 3	Main Steam Line Break (MSLB) type 1
	Large LOCA type 1

- containment isolation,
- stopping the progression of the CD,
- control the hydrogen and monoxide carbon generation to the containment,
- containment integrity for severe pressure and temperature challenges,
- venting and make up of the containment and long-term heat removal, chain control in the reactor and the physico-chemical interaction of core melt and concrete and reactor vessel.

In Fig. 3.3, a sample list of grouped end states in a CET is provided. It includes the following:

The probabilities of sequences leading to an 'OK' state:

$$Pr(Seq_1) = q_0 \cdot \prod_{i=1}^{4}(1 - q_i) \tag{3.1}$$

$$Pr(Seq_2) = q_0 \cdot q_4 \cdot (1 - q_5) \cdot \prod_{i=1}^{3}(1 - q_i) \tag{3.2}$$

$$Pr(Seq_5) = q_0 \cdot (1 - q_1) \cdot q_2 \cdot (1 - q_3) \cdot (1 - q_4) \tag{3.3}$$

$$Pr(Seq_6) = q_0 \cdot (1 - q_1) \cdot q_2 \cdot (1 - q_3) \cdot q_4 \cdot (1 - q_5) \tag{3.4}$$

The probability of the sequence Seq_9 leading to the level of very low Release Category (RC0):

$$Pr(Seq_9) = q_0 \cdot q_1 \tag{3.5}$$

The probabilities of the sequences leading to the low level of release (RC1):

$$Pr(Seq_7) = q_0 \cdot (1 - q_1) \cdot q_2 \cdot (1 - q_3) \cdot q_4 \cdot q_5 \tag{3.6}$$

$$Pr(Seq_8) = q_0 \cdot (1 - q_1) \cdot q_2 \cdot q_3 \tag{3.7}$$

The probability of the sequence Seq_4 leading to the medium level of release (RC2):

$$Pr(Seq_4) = q_0 \cdot \prod_{i=1}^{3}(1 - q_i) \tag{3.8}$$

The probability of the sequence Seq_3 leading to the highest level of release (RC3):

$$Pr(Seq_3) = q_0 \cdot q_4 \cdot q_5 \prod_{i=1}^{3}(1 - q_i) \tag{3.9}$$

It's worth to mention that the SUA task for the PSA level 2 is very important and has significant difficulties, due to the specific features of the deterministic–probabilistic combination. However, the methodology is basically the one presented in the previous paragraph with examples from PSA level 1. The level 2 PSA is performed both for IPSA and EPSA models.

Example 2 solution $\mathbf{S}_{10}^{\mathbf{PR}_{10}}$: There are cases of NPP, especially for the generation IV reactors for which there is no core melt either, because

- it is already melt (the fuel and the moderator are flowing in a liquid form through the reactor and to the thermodynamic cycle—usually two or three thermodynamic heat removal cycles) or because
- there is no core melt, as in some gas reactors, in which the thermodynamic cycle is of very high efficiency (Brayton cycle).

As mentioned in Example 1 solution $\mathbf{S}_{4}^{\mathbf{PR}_{4}}$, the specific of such NPPs is that there is no CD, as the fuel elements will get cracked under high temperature and/or pressure and/or other external conditions, without influencing and generating the mass propagation to other elements (which are of the order of hundreds or thousands) (Fig. 3.4).

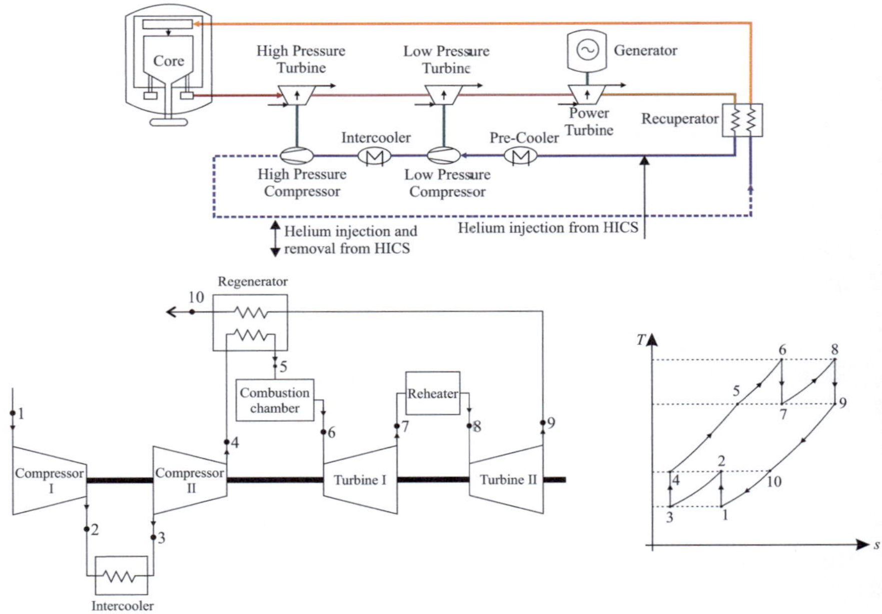

Fig. 3.4 Sample of an NPP with one Brayton cycle

The propagation of the fuel element failures from one to another is one of the main characteristics of the Core Damage (CD) propagation, which is missing in such plants. The result is that the NPP may experience direct releases of various magnitudes, i.e. the PSA level 1 and PSA level 2 are getting combined in a single model, with the ES under the form of the RC.

The limits imposed to the postulated events are defined in a specific manner. For instance, the events might be classified depending on the impact on risk and their frequency (Fig. 3.5) [1–3].

The definition of postulated events leads also (mainly in such cases) to reopening the debate on the Defence in Depth (DiD) layers, i.e.

- how independent they are, and
- how to model them.

There is a research in this direction interfering also with PSA models and some of its aspects are presented in the research activities paragraph.

Figure 3.6 represents the flow path of the PSA level 1 to 3 for an NPP in which there is no core melt [1–4]. There are some specific features to be mentioned about the specifics of these tasks:

- The flow path of the PSA is depending in a much higher degree than in old designs on the computer models (CFD models).

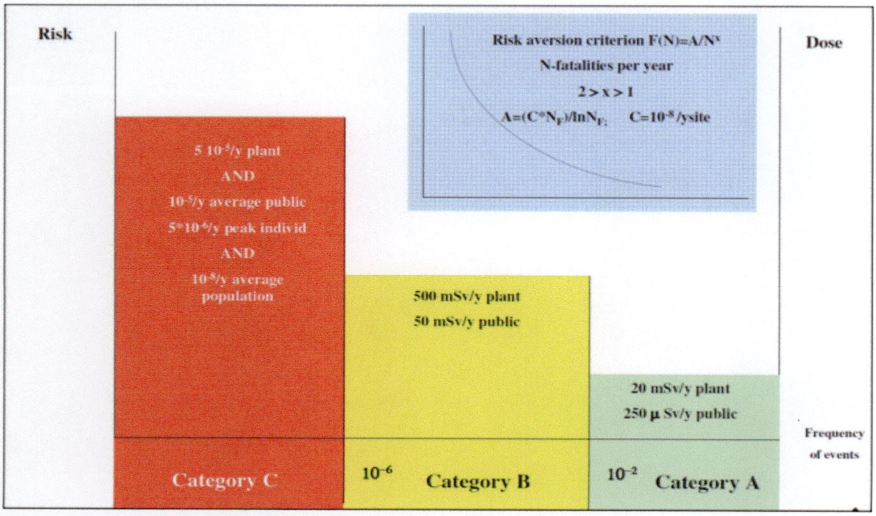

Fig. 3.5 Sample of limits to postulated events in generation IV type NPP

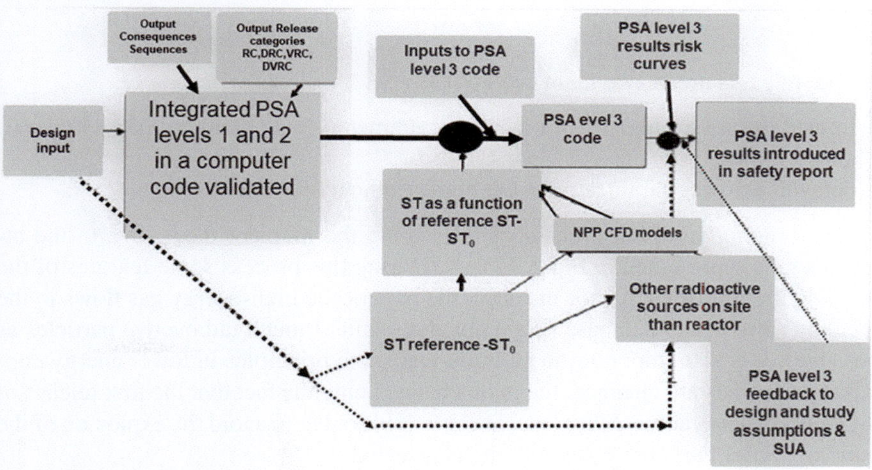

Fig. 3.6 Flow path of PSA tasks (level 1 to 3) in generation IV type NP

- PSA is developed both for licensing purposes and for design optimization, during the design phase.
- The concept of containment is focused rather on the releases than on leak tightness, as the latter is almost impossible to control at the gas pressures and temperatures in case of accidents.

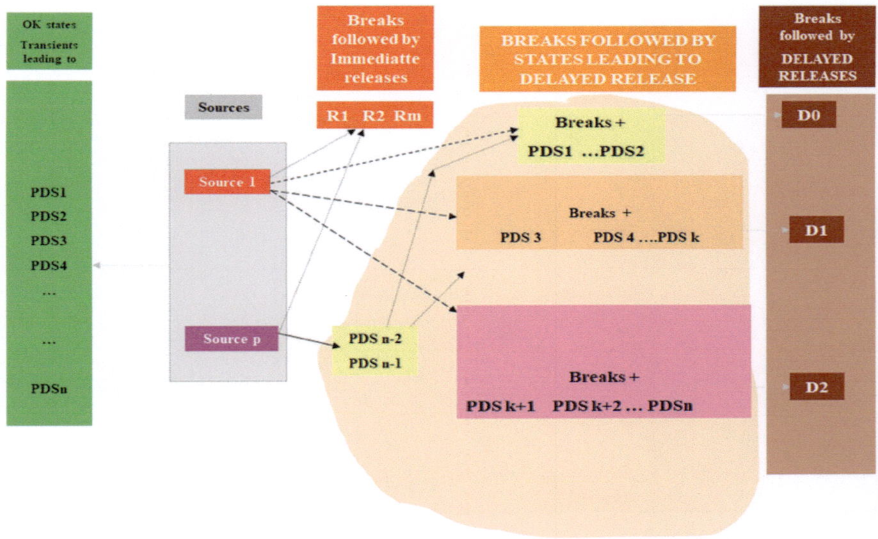

Fig. 3.7 Binning rules of the risk metrics from PSA level 1 to be prepared for PSA level 2 input

The releases are of two main categories:

- immediate at very high thermodynamic parameters, but at low-radioactivity level, and
- delayed at lower parameters, but at higher radioactive releases.

The binning/grouping process of releases to the moment they leave to the atmosphere is represented in Fig. 3.7 [1–4]. During this process some features of the building are modelled, as, for instance, the path of the high-energy gas flows in the building (directed to decrease its energy and retain as much radioactive particles as possible). It is also important to mention that some principles in lower energy containment models are changed, the main change being the fact that the first release of high-energy low-radioactivity level is controlled so that to avoid the explosion of the building. After the release the building is resealed.

The result of the CET task is a set of ES, which are of the following type:

- OK states, leading to failure of some process systems, without risk of releases of radioactivity,
- Immediate releases (coded 'RC i'),
- Delayed releases (coded 'DRCj'),
- Connecting states with other ET of the model.

The results are represented in sample CET in Figs. 3.8, 3.9, 3.10 and 3.11 [1–4].

Figure 3.11 is illustrating in the best manner the fact that the paths of the flow within the building and the interconnection of the systems lead to a high degree of connection between the ET and the need for adequate code modelling, requiring the intensive use of markers, SF and Switches [1, 2, 4].

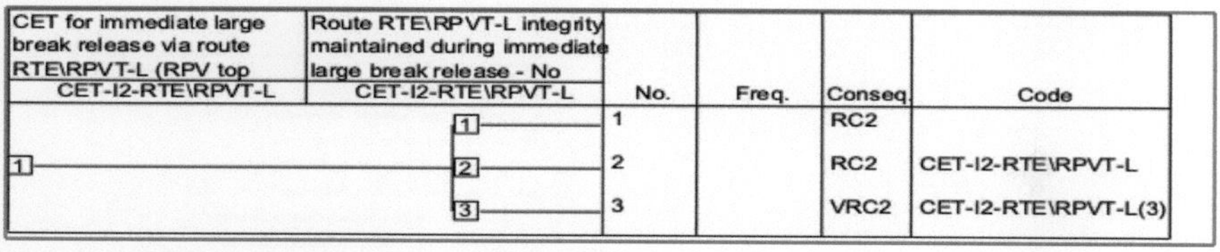

Fig. 3.8 Sample CET for a gas NPP of generation IV:1

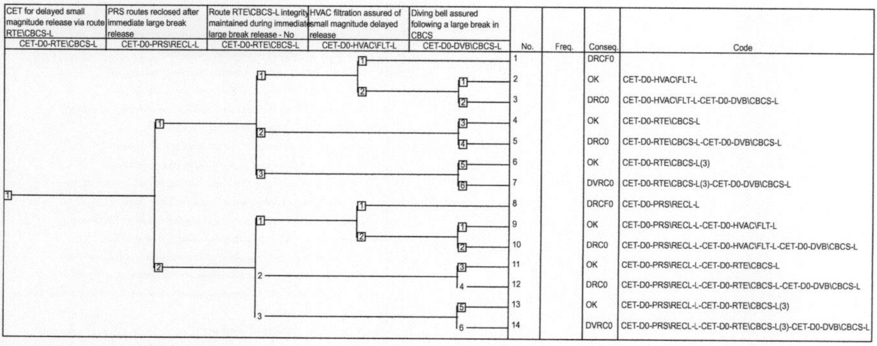

Fig. 3.9 Sample CET for a gas NPP of generation IV:2

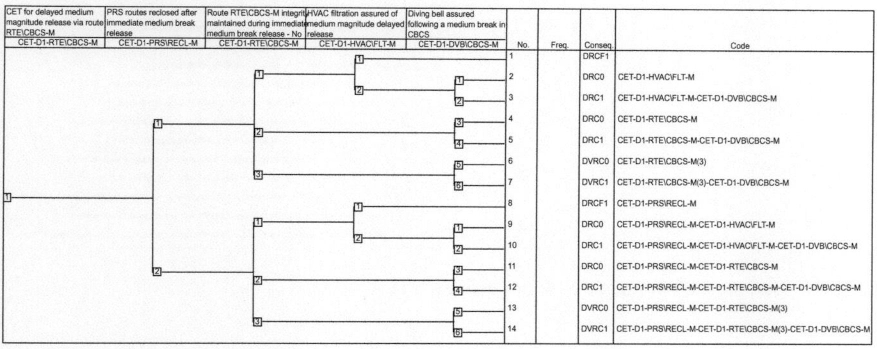

Fig. 3.10 Sample CET for a gas NPP of generation IV:3

The techniques used for the implementation of these needs into the model are similar to those described in previous examples (for instance, in Example 2, Solution $\mathbf{S}_7^{\mathbf{PR}_7}$).

The results of the release evaluations are, as shown in Fig. 3.6, the inputs to the PSA level 3. The main process of the PSA level 3 is to perform a summarization of all the releases for a given distance around the plant and then to calculate the total risk due to possible fatalities.

The goal of the calculations for PSA levels 1–3 for the model I_IPSA-EPSA of such NPP is twofold:

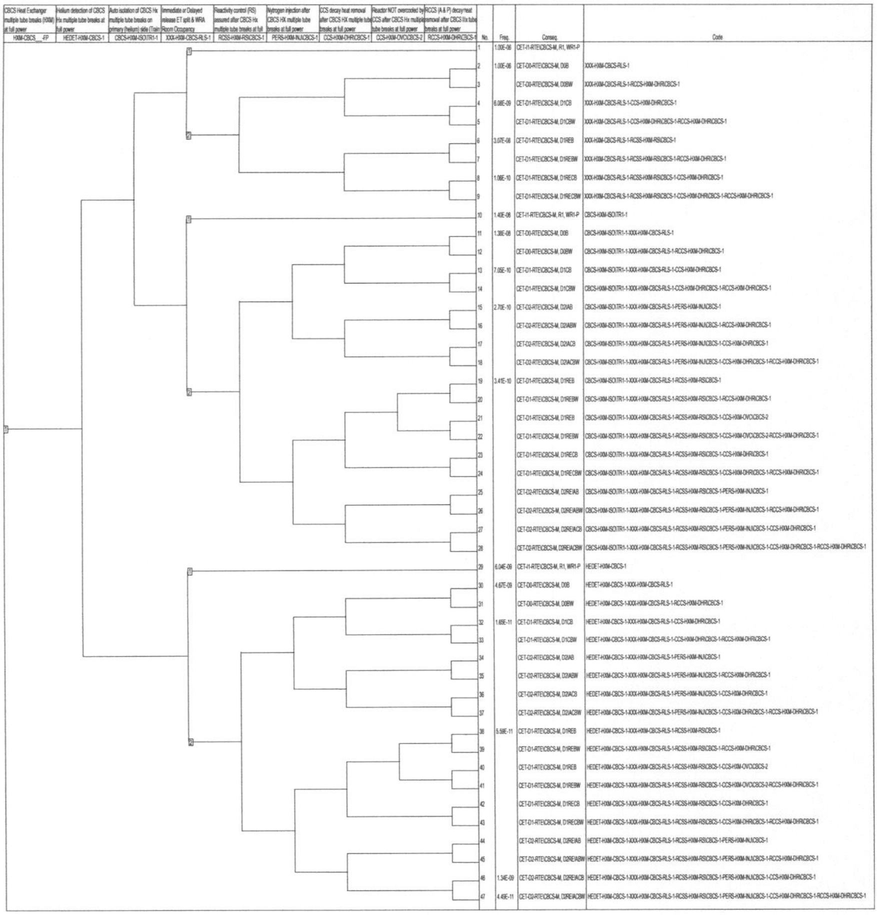

Fig. 3.11 Sample CET for a gas NPP of generation IV:4

- To demonstrate that licensing requirements as the sample in Fig. 3.5 are met;
- To support design optimization from risk perspective during the design phase.

Example 3 solution $\mathbf{S}_{10}^{\mathbf{PR}_{10}}$. HRA modelling in PSA level 2 is a special topic, under intensive research at this moment. The Human and Organizational Factors (HOF) modelling became also a cornerstone of new safety paradigms after Fukushima.

The following aspects are important for HRA modelling in PSA level 2 in order to review the operator model for severe accidents cases, as follows:

- There are levels of difficulty in performing operator actions, requiring more detailed modelling.

  The levels are the following:

  - Low, for actions with a time window of several hours, performed based on clear and written guidance, as, for instance, unblock the containment filtered ventilation in the long term.
  - Medium, for actions with time window of about half an hour, to be performed based on clear and written guidance, as, for instance, the actions in the severe accidents procedures.
  - High, for actions with a time window of minutes and without clear written procedures, for which recovery actions need to be are further defined for emergency procedures.

- The dependencies between the different human actions modelled in Level 2 CETs need to be evaluated as follows:

  - No dependency between the actions included in Level 2 PSA model and those included in Level 1 PSA model. It is usually assumed that the actions modelled as part of Level 1 models are performed by the MCR/SCA crew in accordance with the IR instructions while the actions modelled in Level 2 PSA are performed by the emergency staff in accordance with the SAMG provisions;
  - After any preceding human error in the Containment Event Tree (CET), the operator actions are set to one difficulty category higher than assigned without a preceding human error;
  - After any two preceding human errors in the containment event tree, the operator actions are set to two difficulty categories higher than assigned without a preceding human error;
  - If the preceding operator action was successful the same difficulty category will be maintained as for the case without a preceding operator action.

It is assumed that the emergency staff will be well trained in the Severe Accident Management Guidelines (SAMGs) and associated enabling procedures, similar to that of emergency operating procedure.

The HRA modelling in a systematic manner, able to be integrated into the PSA model is also a target of the research.

Example 4 solution $\mathbf{S}_{10}^{\mathbf{PR}_{10}}$. One important application of the results from PSA level 2 is to support the development of the technical basis for the Emergency Planning (EP). The connection may be done directly from the PSA model, as shown in Fig. 3.12 [5], in a similar manner as performed for transition from level 1 to level 2 (as per Figs. 2.30 and 2.31, for instance).

On the other side, there are specific requirements for the development of an EP in an NPP, as, for instance, the requirement for the Decision Trees DT formulated in (Fig. 3.13 and Table 3.1) [5]. The DT end in Decision States (DS) similar to the ES in PSA models (Fig. 3.13 and Table 3.1).

The connection between the PSA results and the Emergency technical basis evaluations might be done by integrating the model of the PSA into the decision trees described above.

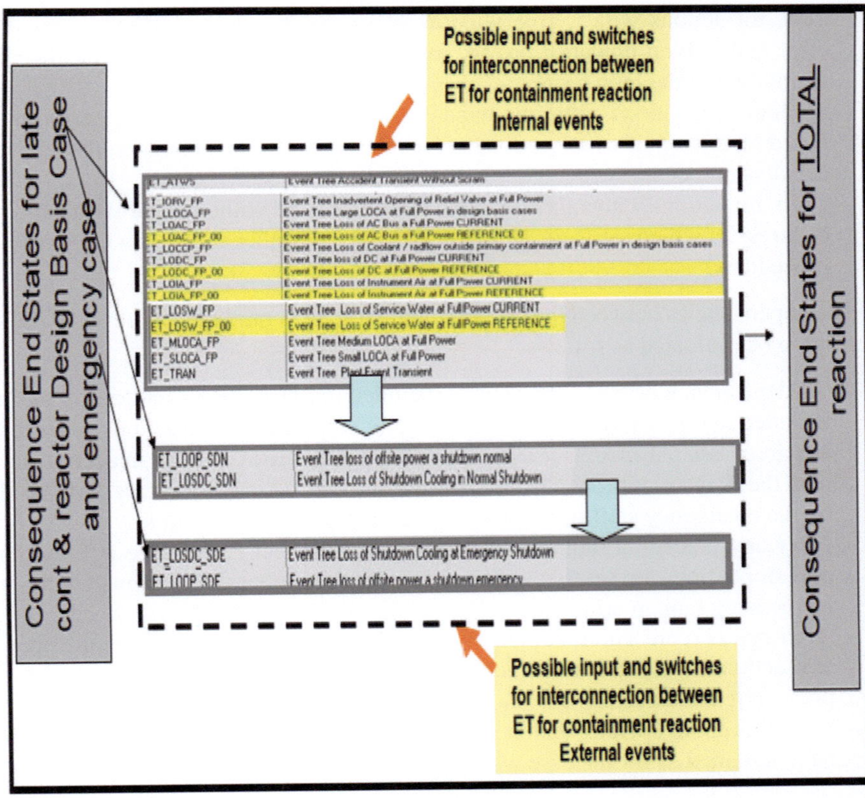

Fig. 3.12 Build internal events model reactor part reaction for the emergency case

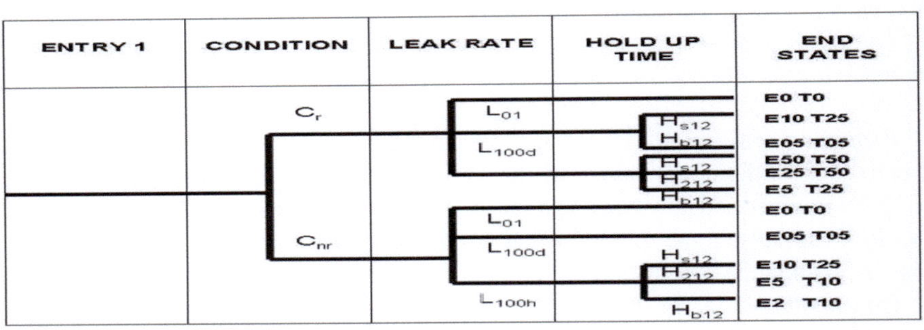

Fig. 3.13 Decision tree for an entry to a scenario leading to various levels of emergency

Table 3.1 Uncertainty and ranking of emergency trees scenarios

Cr	Cnr	L01	L100d	L100h	Hs12	H212	Hb12	Uncertainty	Decision Emergency End States	Final Rank of DES
								Low	E0 T0	I
								Low	E0 T0	I
								Low	E05 T05	II
								Low	E05 T05	II
								Medium	E2 T10	III
								Medium	E2 T10	III
								Medium	E5 T25	IV
								High	E25 T50	V
								High	E50 T50	V

For a given entry the emergency states that include the inputs of end states as defined in PSA level 2 and results from PSA level 3 might be represented and grouped as in Table 3.1. The entries into the DT are derived from the various inputs, including the PSA level 2 and 3 results. The IE of various types and information from other sources (deterministic analyses, OPEX, etc.) are included in the initial list (IEEi). These events are grouped based on three features, as events of

- Symptom-based type (SDT),
- Boundary type (BDT),
- Event type (EDT).

The obtained categories are then grouped depending on the ES from the DT (type of EP status: Alarm, Site Emergency, etc.). Further combination using the DT modelled as logic combinations of failures of barriers that could lead to the final plant and site status from EP point of view. The process is represented in principle in Fig. 3.14 [5].

Further modelling is performed (Fig. 3.15) for the DT in the format of ET, in which the FE are the decisions on (point of decisions are mentioned in Fig. 3.13 and combinations for a generic case are in Table 3.1) [5]:

- The fulfillment of entry conditions into a certain process for EP.
- The status and degradation of the core.
- The leak tightness of the containment.
- The status of the timing in holding up the releases before leaving the containment.

The results of the DT are a list of combinations for each ES in the DT, which are DS. The decision states are actually the emergency categories. The results are in the format shown in Fig. 3.16.

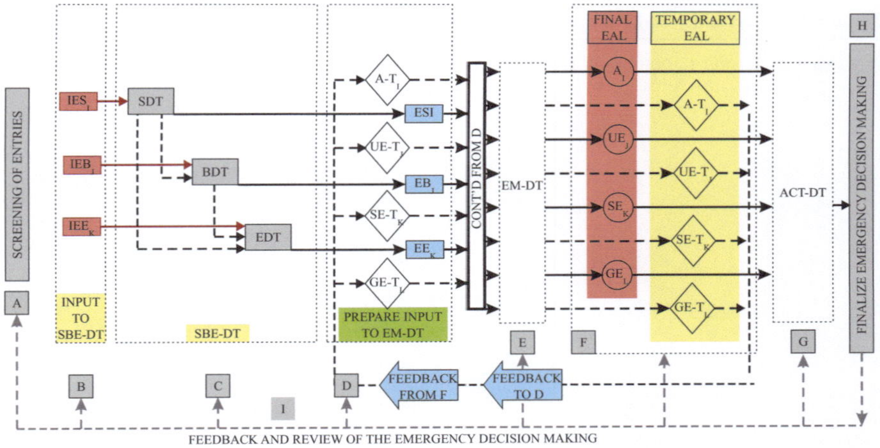

Fig. 3.14 Flow path of connecting PSA level 1 and 2 results with the decision trees for technical basis of the emergency plan (1)

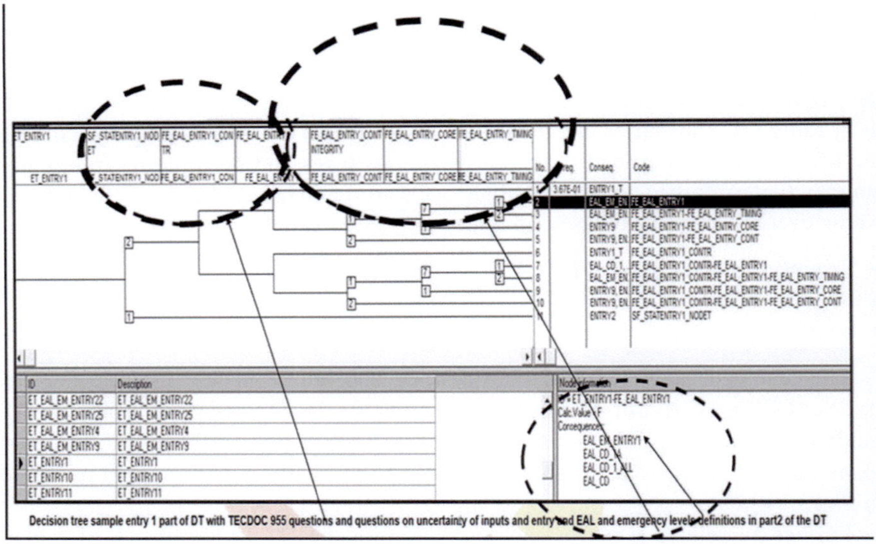

Fig. 3.15 Flow path of connecting PSA level 1 and 2 results with the decision trees for technical basis of the emergency plan (2)

The DS are called Emergency Action Levels (EAL). Further combination of the PSA level 2 results as inputs into IEEi in Fig. 3.14 assure the integration of PSA levels 1 and 2 with the Technical basis of the EP. This model has many advantages, of which the most important are related to their traceability.

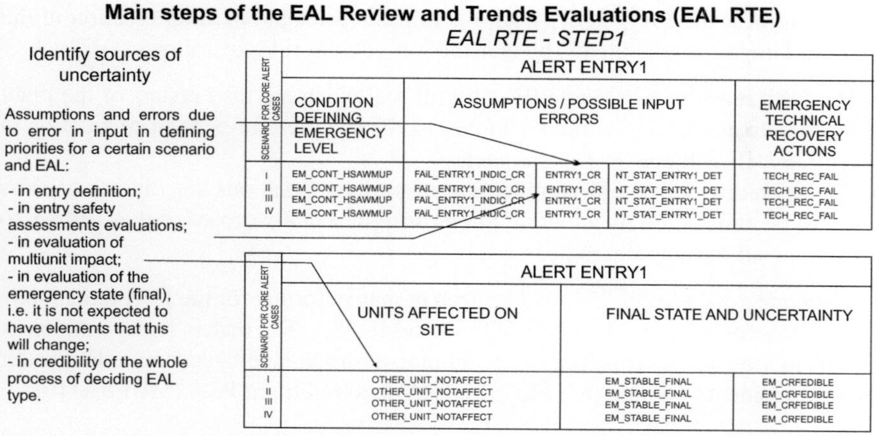

Fig. 3.16 Flow path of connecting PSA level 1 and 2 results with the decision trees for technical basis of the emergency plan (3)

Example 5 solution $\mathbf{S}_{10}^{\mathbf{PR}_{10}}$. SAMG may be modeled also as a Decision Tree (DT), which are connected with the PSA level 2. The combinations for SAMG steps are in MCS format (Fig. 3.23) and are obtained from an SAMG ET model (Fig. 3.18). The SAMG ET include the following:

- Entry into a phase of SAMG,
- Actions to succeed in the steps of the SAMG, for instance, in the Containment pressure Diagnosis,
- Next decision point in the SAMG,
- Exist from the scenario.

Some general aspects considered here are

- The site risk metrics is yet an aspect of debate and left for this moment at the level of decision of nuclear regulators. However, from a technical point of view the only level where risk metrics has a meaning is for PSA level 3;
- For the other levels various rules for CDF and LERF are a matter of high debate and, in our opinion actually some of them (CDF) do not have a physical sense; therefore a MUPSA has to be performed at level 3, so that to have meaningful risk metrics.

Multiunit PSA is one of the safety paradigm changes in PSA approach after Fukushima accident. At this moment in time, there are no validated and standardized methods for MUPSA. However, the results and solutions for some problems presented here are getting more and more confirmation in PSA community.

The Key Topic for the Multiunit ($\mathbf{KT}_{11}^{DMP}$) – $MUPSA$ is related to the techniques to be used in order to connect the specifics of multiunit to the SUPSA model.

The problem for the $\mathbf{KT}_{11}^{DMP}$ ($\mathbf{PR}_{11}$ $\mathbf{KT}_{11}$) is how to implement the specific aspects of multiunit IE on the NPP and to connect MUPSA to SUPSA.

Solution for the $\mathbf{PR}_{11}^{\mathbf{KT}_{11}}$ (S_{11} $\mathbf{PR}_{11}$) is to use a principle of simplification of the multiunit impact on the plants, by considering guiding rules as follows:

- The NPP reaction to a single IE will indicate which are the barriers of the plant reaction to any IE (including the multiunit IE) and therefore, the plant reaction to multiunit IE will consider this aspect;
- The operating states and combination of various situations for diverse sources are to be considered by adopting the most representative cases, rather and try to evaluate all the combinations.

Example 3 solution $\mathbf{S}_{11}^{\mathbf{PR}_{11}}$: An approach in matrix format for the PSA model may be the basis for transition from the SUPSA to MUPSA. The matrix representation of the PSA model (3.10) is helping in generating a format able to accommodate better the development of single-unit PSA (SUPSA) into multiunit PSA (MUPSA) [6].

$$\begin{bmatrix} IE_1 \\ IE_2 \\ IE_3 \\ \vdots \\ IE_n \end{bmatrix} \otimes \begin{bmatrix} c_1 \\ c_2 \\ c_3 \\ \vdots \\ c_n \end{bmatrix} \otimes \begin{bmatrix} a_{11} & \ldots & a_{1n} \\ a_{21} & \ldots & a_{2n} \\ \vdots & \ldots & \vdots \\ a_{n1} & \ldots & a_{nn} \end{bmatrix} = \begin{bmatrix} S_1^1 \\ S_2^1 \\ S_3^1 \\ \vdots \\ S_n^1 \end{bmatrix} \xrightarrow{!S_k^j} \begin{bmatrix} SeSq_1^1 = !S_1^1 \\ SeSq_2^1 = !S_2^1 \\ SeSq_3^1 = !S_3^1 \\ \vdots \\ SeSq_n^1 = !S_n^1 \end{bmatrix} \tag{3.10}$$

IE	CORRELATION FACTORS SINGLE UNIT				CORR FACTORS MULTIUNIT	FAILURE SINGLE AND MULTI		
IE_ISU_3_1	COR3_OPS	COR3_SDS				MH_SDS_3	OPS_U3	
IE_ISU_2_1	COR2L_ISU_IE	COR2_OPS	COR2_SDS			MH_SDS_2	OPS_U2	
IE_ISU_1_1	COR1L_ISU_IE	COR1_OPS	COR1_SDS			MH_SDS_1	OPS_U1	
IE_ISU_3_1	COR3_OPS	COR3_SDS				EP_SDS_3	OPS_U3	
IE_ISU_3_1	COR3_OPS	COR3_SDS				IC_SDS_3	OPS_U3	
IE_ISU_1_1	COR1L_ISU_IE	COR1_OPS	COR1_SDS			EP_SDS_1	OPS_U1	
IE_ISU_2_1	COR2L_ISU_IE	COR2_OPS	COR2_SDS			EP_SDS_2	OPS_U2	
IE_ISU_2_1	COR2L_ISU_IE	COR2_OPS	COR2_SDS			IC_SDS_2	OPS_U2	
IE_ISU_1_1	COR1L_ISU_IE	COR1_OPS	COR1_SDS			IC_SDS_1	OPS_U1	
IE_ISU_3_1	COR3_OPS	COR3_SDS				MH_SDS-ALL	OPS_U3	
IE_ISU_2_1	COR2L_ISU_IE	COR2_OPS	COR2_SDS			MH_SDS-ALL	OPS_U2	
IE_EMU_1	COR1_OPS	COR1_SDS	COR2_EPS		CORR_IE_EMU_OTHER_H	M1\EPS_COM	OPS_U2	
IE_EMU_1	COR1_SDS	COR2_EPS	COR2_OPS	COR2_SDS	CORR_IE_EMU_OTHER_H	M1\EPS_COM	OPS_U1	OPS_U3

Fig. 3.17 Sample result of MUPSA model as an input to the PSA matrix modelling

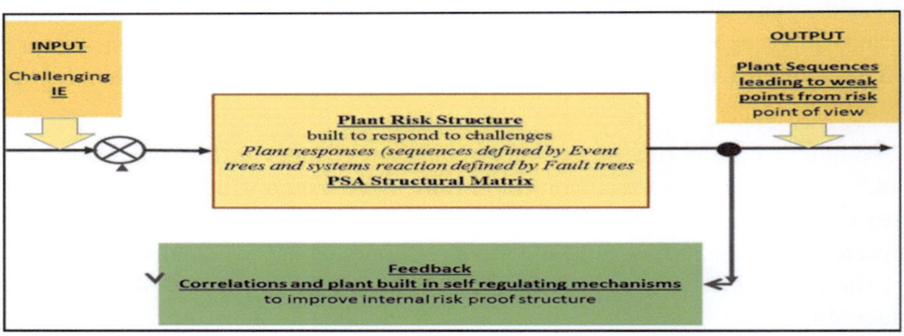

Fig. 3.18 PSA model developed for an NPP that is represented as a cybernetic machine

The presentation of the PSA model in the format (3.10) is based on results of the type represented in Fig. 3.17 [6, 7].

The results that consider both formula of (3.10) type and Fig. 3.4 illustrate the fact that the MUPSA and SUPSA models represent a cybernetic type of plant reaction to risk challenges, as shown in Fig. 3.18 [3, 6, 7].

This reaction may be represented as a 3-dimensional (3D) of a model for MUPSA, SUPSA and the connecting parts between them.

This concept allows a better post processing and interpretation of results, from the point of view of identifying the impact of multiunit effect on the PSA model.

In this case the PSA model (based on formula (3.10) is of 3-D (Fig. 3.25)).

The following SUPSA elements are defined:

- groups of one common failure,
- common cause failure elements (CCF),
- HE of recovery type for single unit.

It is worth to mention that the 'e' components are basic event failures and 'HE' events are human errors. Also, 'k' is a parameter indicating the level of impact of the plant structure on the 3D_RED and it takes a value in the interval from 0.0001 to 2, 'k' and the other parameters listed above being the subject to the parametric sensibility evaluation (Figs. 3.19 and 3.20).

Readers are referred to [6, 8] for complete references.

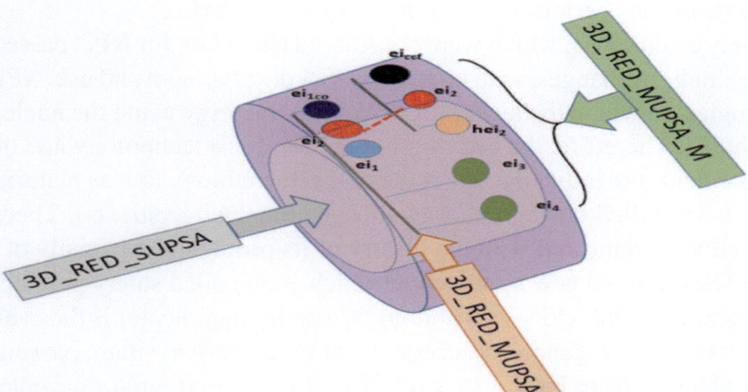

Fig. 3.19 3D MUPSA model representation in a parametric 3D approach (1)

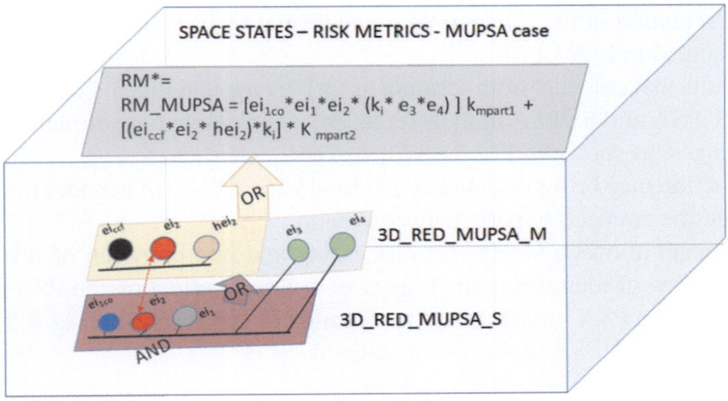

Fig. 3.20 3D MUPSA model representation in a parametric 3D approach (2)

3.1 Use of PSA Results

- PSA and the safety paradigms,
- Use of PSA results in applications,
- Use of PSA results in the decision-making process.

The use of the PSA results is part of the general safety evaluations for an NPP, dominated in various periods by a set of safety paradigms.

The safety evaluations, which were considered necessary for NPP, passed through a series of paradigm changes, with impact on PSA development and use. NPP is a sum of technologies, of which the technology to produce energy using the nuclear energy is dominant [3]. Therefore, the evolution of the NPP as a technology and the history of major accidents are indicating the route of this technology to it as maturity and the problems, as identified during major challenges (nuclear accidents). Therefore, the history of NPP is connected with the history of its problems, especially of its major accidents. They defined new approaches, which were called safety paradigms [3, 9, 10]. For a history of the NPP as a technology, a defining indicator is the evaluation of the safety margins (it is generally accepted that the reserve the main parameters with impact on plant damage have a to a set of limits imposed and/or accepted). From this perspective, the risk metrics are some of the indicators of the safety margin. The evolution of NPP technology, as reflected in the safety margins and in the risk metrics may be described by the *s*-curve of a given technology, where s is, in our case, the safety parameter including risk metrics.

3.1.1 PSA and the Safety Paradigms

The history of PSA is tightly connected with the history of safety analyses, highly dependent on the OPEX of NPP, throughout the world, in the past half of the century. The *OPEX* impact is visible for any significant event. Therefore, the major nuclear accidents lead to major changes in the safety analyses and defined new paradigms in which the NPP safety performance was evaluated, as illustrated in Fig. 3.21 [3, 10].

For the PSA history this involved the existence of the following major safety paradigms changes:

- Post TMI accident period was defined as the DiD paradigm in all safety analyses. The generation of the concept and its implementation were connected with the first PSA studies. During this period PSA started to be developed in a systematic manner and basic standards were issued [3, 7, 10]. PSA was considered as a complementary tool to the deterministic analyses and a large series of studies were initiated for many NPPs. A process of implementation at the world scale was started. PSA levels 1 to 3 were developed and the incipient PSA or external events were started. However, there was no systematic use of PSA results in the decision-making and less impact and development of PSA levels 2 and 3.
- Post-Chernobyl accident period was defined by the keyword 'emergency' and an emphasis on PSA level 2 and 3 was during this period. The use of the PSA in risk decisions, called Risk-Informed Decision-Making (RIDM) started to be formalized in standards and documents.

  1. RIDM is an important practical tool to be used in most of the licensing systems, which are risk informed. A risk-informed NPP licensing system is a system considering risk evaluations as complementary to the deterministic ones and the OPEX.

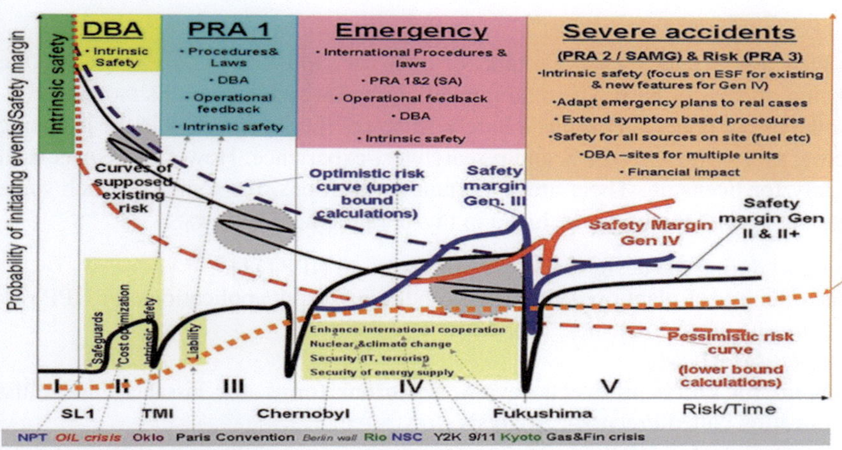

Fig. 3.21 History of NPP safety margins and safety/risk metrics paradigm changes

2. There are cases (as, for instance, in the UK, the Netherlands for Risk-Based Decision-Making (RBDM) in the regulatory process).
3. However, the use of PSA for risk evaluations is not influenced in its methodology by the differences between RIDM and RBDM.

- The present post-Fukushima period is characterized by the paradigm defined on the extension of Design Basis Accidents (DBA) and consideration of the severe accidents and Cliff Edge Effects (CEE). The consideration of multiunit multisource impact is also a new effect of the paradigm changes.

However, there are two aspects to be retained in the light of the topics of this book:

- For all this period, the development and use of PSA revealed a series of aspects, for which support to practitioners is of high practical importance. Some of them are highlighted, with proposed possible solutions in this book;
- Research on the PSA tools never stopped.

During all these periods, in parallel with the standards development, intensive research activities were performed to support PSA methodologies, as, for instance:

- Development and validation of high-performance computer codes for PSA levels 1 to 3.
- Mathematical and logical background of PSA methodologies.
- Research on phenomena for PSA level 2 and development of specialized codes.
- Research on level 3 and use of PSA at all levels for various applications.

Some special issues related to the research activities listed before are also mentioned in this book.

3.1.2 Use of PSA Results in Applications

The PSA levels 1 to 3 are important for their use, which is mainly related to the licensing process of NPP. The requirements agreed worldwide at this moment consider the use of both Deterministic Safety Analyses (DSA) and PSA in the licensing process, aside with the OPEX and research/test experience. However, PSA is used not only for licensing. There are important applications, for some of which some important special topics are included in this paragraph, as follows:

- Support during the design/redesign process of an NPP [11].
- Risk monitor of plant operation (most codes have now applications for NPP risk monitoring, as, for instance, [12, 13]).
- Support for RIDM.
- Support for OPEX in events review in various forms, for instance, under the application called Precursor Analysis.
- Severe accidents modelling and support to the Emergency Planning (EP) Technical Basis.

3.1.3 Use of PSA Results in the Decision-Making Process

The use of PSA to support RIDM is an application, which became an important tool in the licensing process both for licensees and the regulatory organizations.

An example of issues to be considered for this application is presented in some examples below.

The Key Topic for the Use of PSA results ($\mathbf{KT}_{12}^{GR12}$) is to develop techniques to prepare the PSA like results for their use in various applications.

The problem for the $\mathbf{KT}_{12}^{GR12}$ ($\mathbf{PR}\_12^{\mathbf{KT}\_12}$) is twofold:

- The applications are diverse and that the PSA results are not fit for their use for such purposes. As a result, it is necessary to develop new approaches so that, the PSA results are being able to be used in applications.
- PSA levels 1 to 3 PSA has a series of limitations, usually not carefully considered in defining the limits of using them in applications.

The Solution for the $\mathbf{PR}\_12^{\mathbf{KT}\_12}$ ($\mathbf{PR}\_12^{\mathbf{S}\_12}$) is to solve the two main challenges of PSA use in applications:

- Build special adapting tools for the use of PSA results in specific applications.
- Increase the level of understanding of the PSA results limitations.

Example 1 solution $\mathbf{S}_{12}^{\mathbf{PR}_{12}}$: PSA paradigms and limits.
PSA limitations are coming mainly from the manner the method is build as a combination of

- PSA is a set of logical combinations using the Boolean algebra rules for describing the possible scenarios that could lead to the NPP critical situation from the risk metrics point of view [14–16]. However, the assumptions are subject to extensive reviews, called Sensitivity Analyses, which have the goal to define in detail the limitations.
- PSA is based on results from DSA and OPEX and therefore there is a tight connection, but with clear areas of applicability, between PSA areas of applicability and DSA [14, 15].
- The PSA technique assumes that the behaviour of its elements (plant equipment, systems, hazards challenging NPP, etc.) are of probabilistic nature and have a certain distribution. Therefore, a combination of probabilistic elements is subject to the rules of combination of probability theory. Even if the results are presented in mean values the uncertainty bounds are a matter of special detailed calculations.
- The probabilistic type of reasoning is to be used in the evaluation of the results.

This example is presenting an approach, which is able to give answers to the issues mentioned above. PSA has a well-defined area of applicability and there is a certain type of safety issues where it is best recommended, in the context of diverse methods [16] for the safety evaluation (as represented in Fig. 3.22).

The NPP safety evaluation is using diverse tools, as such:

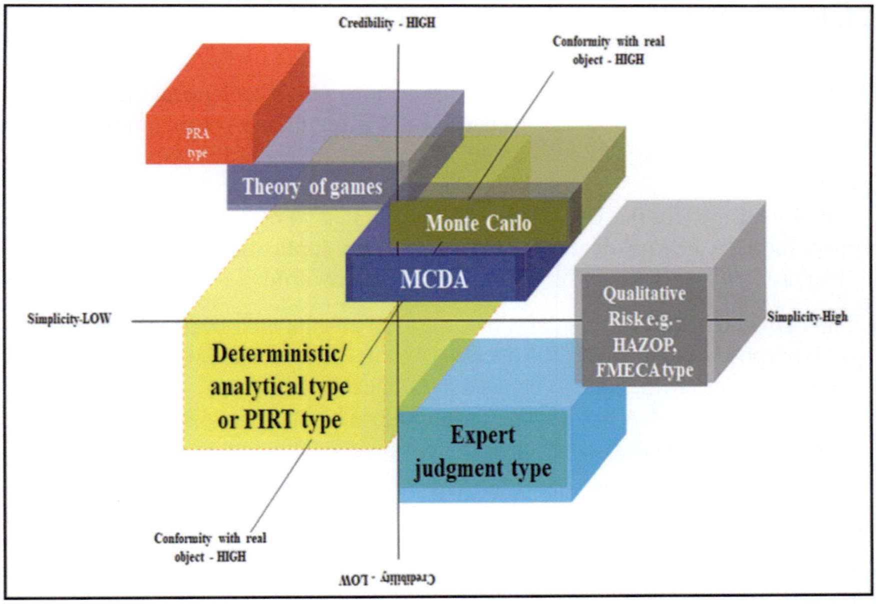

Fig. 3.22 A set of methods available in the toolbox of safety analyses

- PSA,
- DSA, including a combination of expert and DSA (as PIRT and SOARCA in US NRC) [17, 18]
- Theory of games,
- MCDA (Multi-criteria Decision Analysis),
- Hazard Analysis (HAZOP),
- Failure Mode and Consequence Analysis (FMECA),
- Expert Judgments,
- Monte Carlo modelling and various statistical methods, part of OPEX.

They are important from various perspectives, of which the main are the following three aspects (Fig. 3.23):

- Credibility of results,
- Capability to describe accurate the NPP,
- Level of complexity of the method/tool.

From this perspective, PSA has the following features:

- An area of credibility for its area of applicability, to be detailed further,
- A high accuracy of NPP description,
- Even if the complexity of the method is very high.

For the safety evaluations, various approaches could be adopted. Let us consider the following:

1	AI2054ENDSHDH10	DATB435_RIH6_SUBCOOL	HE_NODIAG	IE_DFC_ENTRY	MCR DOSE	MOD_LEVH6900	MOD_PMP_ON	SF_POWER_FP
2	66611PL5456	DATB432_ROH3_SUBCOOL	HE_NODIAG	IE_DFC_ENTRY	MCR DOSE	MOD_LEVH6900	MOD_PMP_ON	SF_POWER_FP
4	AI2154ENDSHDH10	CCUA DOSE	DATB432_ROH3_SUBCOOL	HE_NODIAG	IE_DFC_ENTRY	MOD_LEVH6900	MOD_PMP_ON	SF_POWER_FP
5	66611PL5456	DATB438_RIH8_SUBCOOL	HE_NODIAG	IE_DFC_ENTRY	MCR DOSE	MOD_LEVH6900	MOD_PMP_ON	SF_POWER_FP
6	AI2054ENDSHDH10	DATB430_ROH1_SUBCOOL	HE_NODIAG	IE_DFC_ENTRY	MCR DOSE	MOD_LEVH6900	MOD_PMP_ON	SF_POWER_FP
8	63411L2728	CCUA DOSE	DATB435_RIH6_SUBCOOL	HE_NODIAG	IE_DFC_ENTRY	MOD_LEVH6900	MOD_PMP_ON	SF_POWER_FP
9	AI2054ENDSHDH10	CCUA DOSE	DATB438_RIH8_SUBCOOL	HE_NODIAG	IE_DFC_ENTRY	MOD_LEVH6900	MOD_PMP_ON	SF_POWER_FP
10	66611PL5456	AI1161_RIHLEVH04	CCUA DOSE	HE_NODIAG	IE_DFC_ENTRY	MOD_LEVH6900	MOD_PMP_ON	SF_POWER_FP
11	66611PL5456	DATB433_RIH4_SUBCOOL	HE_NODIAG	IE_DFC_ENTRY	MCR DOSE	MOD_LEVH6900	MOD_PMP_ON	SF_POWER_FP
15	66611PL5456	CCUA DOSE	DATB436_ROH7_SUBCOOL	HE_NODIAG	IE_DFC_ENTRY	MOD_LEVH6900	MOD_PMP_ON	SF_POWER_FP

Fig. 3.23 The combinations for SAMG steps in MCS format obtained from an SAMG ET model

M1	M2	M3	M4	M5	M6	M7
U	DP	DPR	PRU	PRO	PROU	DPROU
O	PR		DRU	DRO	DPOU	
R	DR		DPU	DPO	DPRU	
P	PU		ROU		DPRO	
D	DU		POU			
	RU		DOU			
	OU					
	PO					
	DO					
	RO					

Fig. 3.24 Combinations of approaches/methods used in safety evaluations of NPP

(i) Deterministic (D),
(ii) Probabilistic (P),
(iii) Operational Feedback (O),
(iv) Quantitative risk analyses (R),
(v) Data, methodology and epistemic uncertainties (U).

The approaches could be used in various combinations, which may be grouped as illustrated in Fig. 3.24.

These combinations presented in Fig. 3.24 are the result of a Decision Tree (DT) shown in Fig. 3.25 [3, 6].

However, the best-known and most used analyses, confirmed by requirements and defined by *NPP standards are DSA and PSA. They are using elements from the other methods and are connected between them* [16, 19].

As far as the dilemma of using deterministic versus probabilistic analyses is concerned, the solution is actually in as accurate as possible definition of the areas of applicability in each case, as illustrated in Fig. 3.26.

Areas BP and A2 are recommended for PSA. They have the following characteristics:

(i) BP, providing also a support to the NPP as CAS Resilience:

- A high level of credibility in the degree of uncertainty of evaluations and hence decisions.
- Even if a low level of credibility in the degree of conservatism of the evaluations.

(ii) A2:

- Acceptable areas of both credibility in the degree of:
- uncertainty of evaluations and hence decisions,
- conservatism of the evaluations.

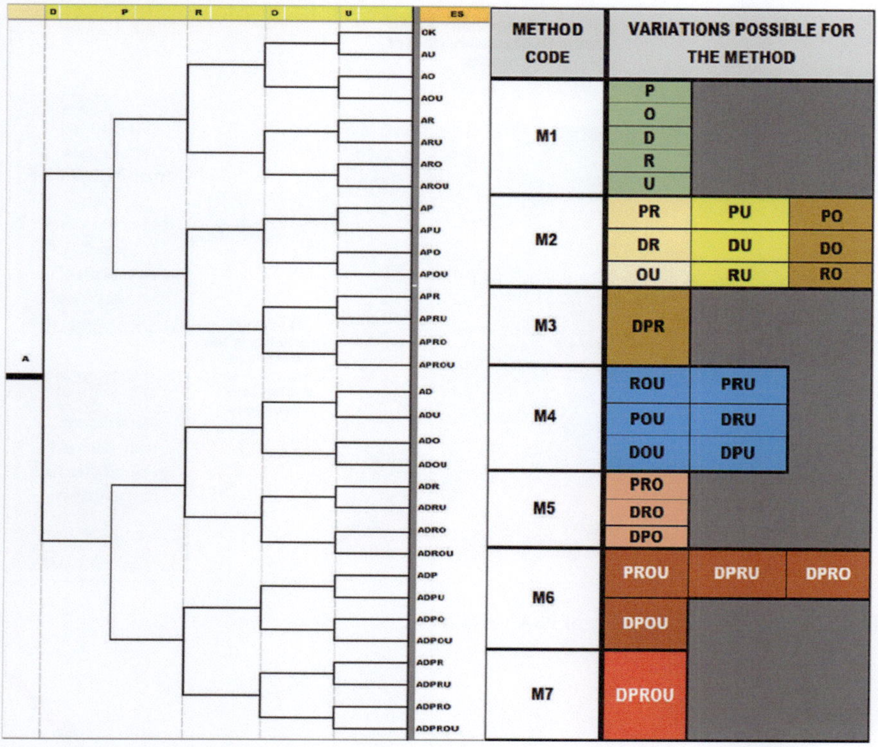

Fig. 3.25 DT for the combination of methods in safety evaluations

For the PSA, the Objectives Function is in correlation with the risk metrics. A simplified representation of the Risk-related objective function and performance Objective functions are in formulas (3.11), (3.12), (3.13) and Figs. 3.27, 3.28 [3].

The total objective function (Y_{TOT}) is a resultant of the optimization of Risk Criterion (RC) and Technology/commercial criterion (TC) defined as per the next formulas and represented in formulas (3.11), (3.12), (3.13).

$$Y_{TOT} = C_0 \cdot e^{c_1 x} + C_2 \cdot e^{x^n} \tag{3.11}$$

$$RC : Y_1 = C_0 \cdot e^{c_1 x} \tag{3.12}$$

$$TC : Y_2 = C_2 \cdot e^{x^n} \tag{3.13}$$

The use of the three main methods during the Decision-Making Process (DMP)—DSA, PSA and OPEX leads to the need to evaluate the total credibility of this decision. In order to perform this task, it is that the decision in the NPP model, considered a CAS, has total Objective Function. Its more general formulation, related not only to PSA as the (3.11), (3.12), (3.13) versions are described for all the analyses in (3.14):

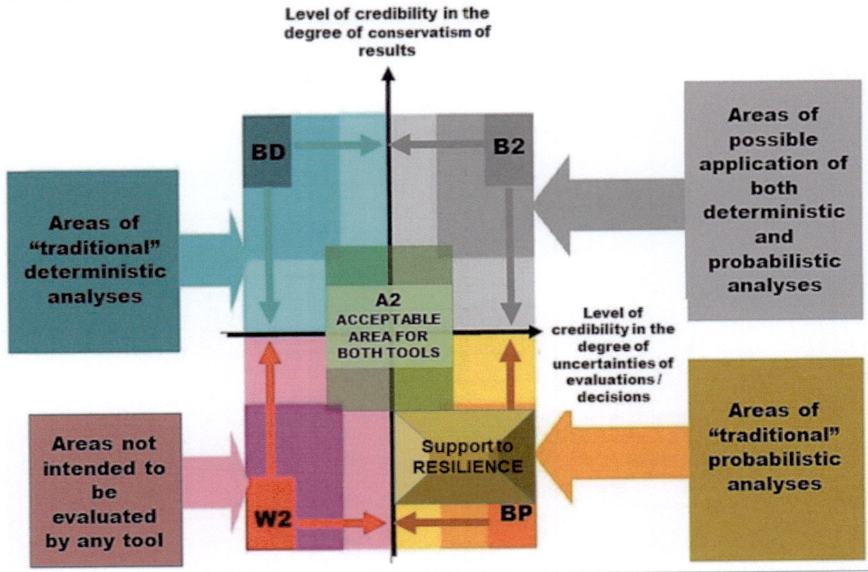

Fig. 3.26 Areas of applicability of PSA versus DSA

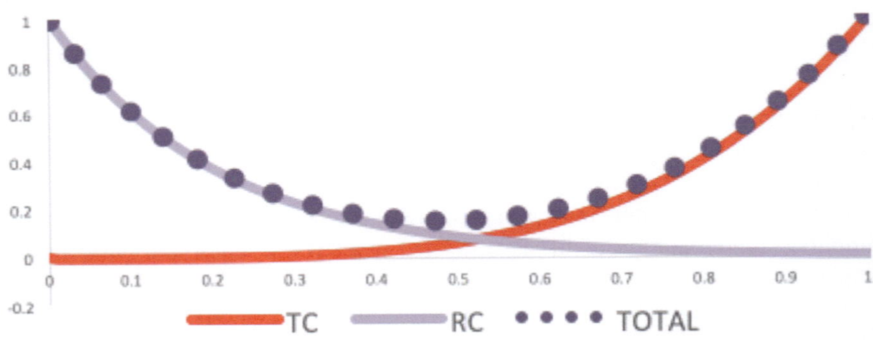

Fig. 3.27 Optimizing NPP objective functions (1)

$$O = (P \otimes_{Rp} U(P)) \otimes_{RG1} (D \otimes_{Rd} U(D)) \otimes_{RG2} (F \otimes_{Rf} U(F)) \tag{3.14}$$

The function O (Objective of the decision process) is a result of a combination using a series of logic operators (Fig. 3.29) [3]:

- RP and D designate reasoning on the credibility of probabilistic and, respectively, deterministic results, and RF for the reasoning on the credibility of reasoning based on feedback from experiments/real cases;
- $RG1$ and $RG2$ for connecting results on reasoning based on the following:

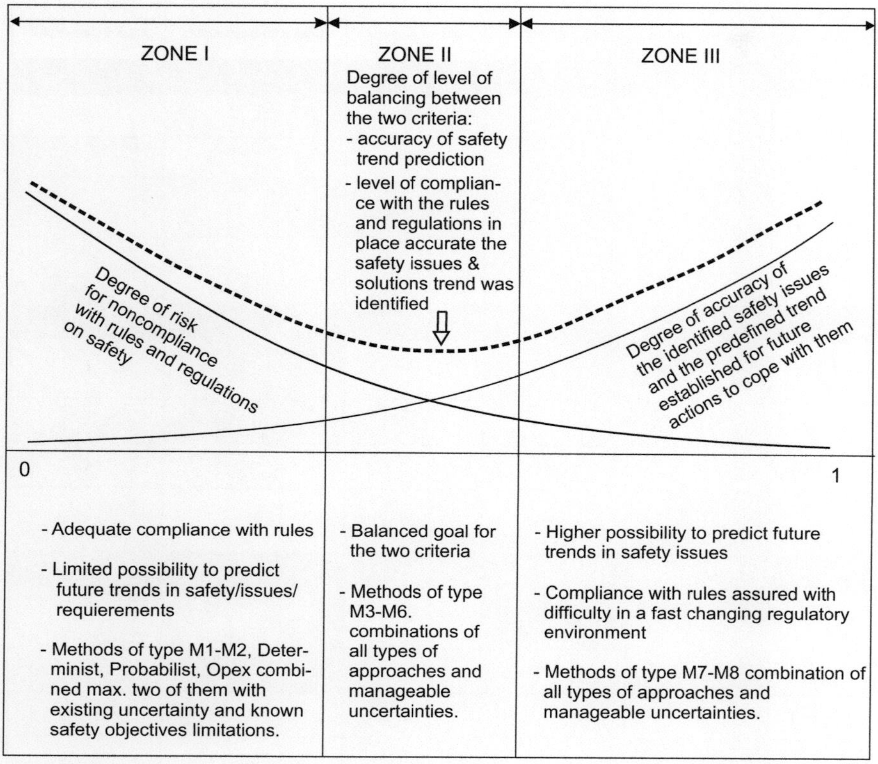

Fig. 3.28 Optimizing NPP objective functions (2)

- probabilistic evaluations (for the terms noted with *P*—probabilistic statements and Up—probabilistic statements uncertainties),
- deterministic evaluations (for the terms noted with *D*—deterministic statements and Ud—uncertainties of deterministic statements),
- feedback review statements (for the terms noted with *F*—statements based on feedback review and Uf—uncertainty of the statements from feedback review).

The decision-making statements are fundamentally divided into 'deterministic'-oriented statements and 'probabilistic'-oriented statements. For deterministic judgments, the result is composed of the criteria value D and the level of uncertainty in these values (Ud); for the probabilistic results, the components of the results are P and Up. There is also a component of results given by feedback from real object while compared to the model (F set of statement).

Operator will have various impacts on the final function (with low—L, medium—M or high—H impact) as shown in Fig. 3.29, depending on the type of judgment cases, in which the decision-maker positions himself (which could be optimistic, pessimistic, etc.). The result of how the final credibility should be considered given

Impact / cases	Case 1 Optimistic	Case 2 Pesimistic	Case 3 Average	Case 4 Overoptimistic	Case 5 Catastrphic
P	L	L	M	L	M
RP	L	L	M	L	H
U(P)	H	H	H	H	H
Total P	L	L	M	L	H
D	H	H	M	H	L
RD	H	L	L	H	L
U(D)	L	L	L	L	L
Total D	H	M	L	M	L
F	H	H	M	H	L
RF	H	L	L	H	L
U(F)	L	L	L	L	L
Total F	H	M	L	M	L
R^G_1	M	M	L	L	L
R^G_2	H	M	M	H	M
O Total Objective Function	M	M	H	L	L

Fig. 3.29 Objective function in various types of DMP

a set of deterministic results is illustrated in Fig. 3.29, which shows that the role of the decision-maker can be also modeled and considered a priori so that variations in the conclusions of the same risk results used by various interest groups could be predicted and understood. Understanding risk results is one of the main conditions of assuring a good risk governance process and maximizing the use and impact of the risk evaluations.

There are fundamental differences between the deterministic and probabilistic approaches. In the first place, they use a different reasoning. In the probabilistic reasoning (formula (3.11)), a statement is of the following type [3]:

'Element X known with uncertainty Ux is requiring element Y known with uncertainty Uy and they are producing a known effect W with uncertainty Uw'.

and this is a significant difference from the deterministic reasoning:

'If X is requiring Y to produce the effect W and the two conditions are fulfilled, then W will take place'.

Therefore, the *use of PSA for decision process* consists of

- *not only in using probabilities instead of average unchanged values/parameters,*
- *but mainly in using a probabilistic type of reasoning.*

These specific aspects lead to a specific set of areas of best applicability for probabilistic approaches (Fig. 3.30), which details the generic representation from Fig. 3.26 [3].

The conclusions on the *DMP using DSA and PSA*, as resulted from Fig. 3.30 are as follows:

- **If the decision is aimed at evaluating high foreseen risk situations above the acceptable limits, then the deterministic pessimistic statements will lead to the most conservative decision, even if that will happen under less credibility**

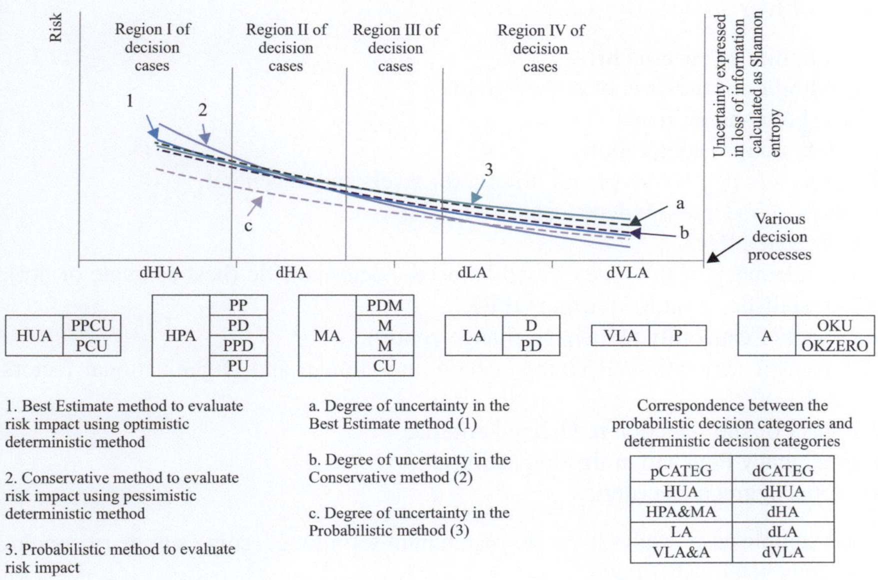

Fig. 3.30 Areas of applicability of PSA from DMP perspective

than for the probabilistic ones. *But by other reasons than technical ones, the deterministic-based decisions could be expected.*

- **If the decision is aimed at evaluating high or moderate foreseen risk situations below the acceptable limits, then there will be no difference between the very pessimistic way of thinking, an optimistic one or a probabilistic one except the fact that the probabilistic one will have more credibility**, *which could make it the most probable choice for the decision.*
- **If the decision is aimed at evaluating low and very low foreseen risk situations below the acceptable limits, then it will be based on the probabilistic approach, giving the fact that it generates the most conservative results with the highest credibility. Evaluation of risk impact using extensive sensitivity cases is one of the key issues to support the probabilistic type of thinking and its more extensive use in decision process**. *This is integrated into the verification and validation process, of which independent review and benchmarking play a very important role in confirming the truth value of probabilistic statements.*

Example 2 solution $\mathbf{S}_{12}^{\mathbf{PR}_{12}}$. Sample case of the use of PSA combined with DSA and OPEX in a nuclear safety set of medium-term evaluations for NPPs. The evaluations also included expert opinion and modelling of Human and Organisational Factors (HOF) [20]. A real care of expert experience of using diverse combinations of safety analyses, in line with the possible approaches mentioned in Example 1 solution $\mathbf{S}_{12}^{\mathbf{PR}_{12}}$ is presented below. The stages of safety evaluations and their features are listed in Figs. 3.31, 3.32 and 3.33 [3, 7]. Detailed criteria for the evaluation of results are used and an evaluation is performed for each phase during a nuclear energy program period of four decades (Figs. 3.31, 3.32 and 3.33).

- Credibility of uncertainties,
- Credibility of the level of conservatism,
- Level of conservatism,
- Safety margin acceptability,
- Defence in depth Acceptance criteria for levels and in general,
- Defence in depth—Independence of levels,
- Cliff edge effects,
- The adequacy of the type of method used—deterministic (best estimate or not), probabilistic, combined, using OPEX,
- Impact of capability to manage change control,
- Impact of generation/technology phase and Human and Organizational factors (HOF),
- Impact of site selection predefined criteria,
- Emergency Plan and mitigating actions,
- Global aggregated criteria.

As sample case results show, in the medium-term range, some summary conclusions may be already drawn:

- *The stability of safety decisions was assured by the complementary of the three types of evaluations: DSA, PSA and OPEX.*

Strategy		Case		Method	
Code	Key elements	Code	Key elements	Code	Key elements
S1	Concepts of SM & DiD are consolidated and recognized internationally in standard like format (period I 1990-2000 in Figure 3.34). Vendors are adapting the initial safety philosophy to the changes in SM and DiD .	BAS-U1	Basic CANDU philosophy as defined by the concept designer to be endorsed and considered as basis for licensing as it is.	M2	Deterministic analyses for a set of Postulated Initiating Events in Final safety Analysis Report and supplementary support documentation of probabilistic analyses (Reliability Analyses RA for some systems and Safety Design Matrices SDM).
		EQUIV RO	Basic CANDU philosophy and a Canadian licensing system non-prescriptive needed to be adapted to a prescriptive regulatory system (implementing 10.CFR 50 and NRC approaches) adopted earlier in Romania for TRIGA.	M2	Licensing meetings considered differences in licensing on an issue-by-issue approach in a regulatory licensing project process. Transfer of regulatory approaches on deterministic and probabilistic tools started with Canadian regulator. Proposals from support from IAEA projects and in late 90's from EU via PHARE projects for Regulatory Body reorganization and norms review implemented.
		PRA 1	Probabilistic approach of basic CANDU of RA and SDM reviewed against PRA level 1 results performed in independent projects under IAEA for Cernavoda Unit1. Results used in combination with EQUIV RO changes in regulatory approach and based on commissioning test results of Unit 1.	M1	Review of the DiD features; review and study possibility to extend DBA category; impact of support systems and the need to consider a higher impact on SM and DiD from their side by comparison with the BAS-U1. Implement design changes as proposed by licensee based on latest CANDU developments (because the Canadian BAS – U1 evolved for other projects).

Fig. 3.31 Strategies and methods used in the evaluated cases (1)

- It is also shown that the impact and role for basic licensing, but also for design optimization and other operating and emergency applications increases for PSA.
- It is also important to mention that the major risk envisaged for the period 4 (next 10 years) is not the fact that post-Fukushima actions will be not implemented, but the fact that
  - Either the change control, i.e. the planning of introducing all those modifications are not functioning,

S2	Concepts of SM & DiD were consolidated. Special issues under research for advanced new generations of NPP and / or SM & DiD problems in the context of lifetime extension issues. Consolidation considered a certain optimism ("nuclear renaissance") and did not anticipate intense actions to review approaches on SM & DiD after Fukushima accident (period II 2000-2011 in Figure 3.34).	U2	Consolidated approach adopted for Cernavoda NPP unit 1 with the lessons learnt and supplementary changes proposed after experience in other CANDU 6 projects completed between 2000 and 2007. No change in SM and DiD requirements. PSA level 1 requirements included as mandatory and requirement to develop PSA level 2.	M2 MOD	Review of the DiD features; review and study possibility to extend DBA category; impact of support systems and the need to consider a higher impact on SM and DiD from their side by comparison with the BAS-U1. Implemented more design changes as proposed by licensee based on latest CANDU developments and included PSA level 1 (internal and external events) in the licensing documentation. RA and SDM kept as orientative indicator of the basic safety design. Started preparation for PSA level 2 and severe accidents evaluations, as well as for ageing impact for long-term operation. Periodical Safety Review (PSR) completed. Risk Informed Decision Making (RIDM) elements started to be implemented
		PBMR	Generation III+ NPP project considering the latest requirements for generation IV. Use of results in RIDM	M3 P	Deterministic and probabilistic evaluations. Risk goals used based on PSA level 1 – 3 evaluations.
		AGE	Review of existing status of the probabilistic methods for the evaluation of the impact of ageing on plant safety in the framework of an EU Ageing PSA network (including development of methods).	M4 A	Develop probabilistic methods to be included in the safety documentation for a more accurate description of the SM if the ageing effects are considered. The project connected with the definition of existing SM after considering ageing and how the levels of DiD are affected. Use of results in RIDM

Fig. 3.32 Strategies and methods used in the evaluated cases (2)

– Alternatively, there is a (possibly) hidden impact (not evaluated in sufficient detail) of those modifications on already existent safety features from the 'traditional' DBA, which may lead to totally unexpected major accidents.

Those possible accidents might be generated by a cavalcade of modifications on designs that do not support them, making things worst, or due to the loss of change control itself. It might be that, by avoiding *Cliff Edge Effect, basic safety feature already existent are challenged and that is why the implementation of changes of post-Fukushima type should consider with priority this aspect and the fact that a certain NPP generation has its safety margin limits* (Fig. 3.34).

More refined modelling of the NPP safety analyses considering the HOF in the context of safety paradigm changes have a continuous impact on the PSA approach. New novel integrative approaches for an I _IPSA _EPSA model considering HOF (in a theory on topological model of NPP as CAS) are under development.

S3	Concepts of SM & DiD under review due to the need to consider extension of the Design Basis Accidents (DBA) in the format of Design Extended Conditions (DEC). DiD under scrutiny. Post Fukushima actions under implementation (Period III 2011-2017 in Figure 3.34).	L2 U1&2	All the requirements on SM and DiD from case U2 valid and in addition PSA level 2 performed for Cernavoda U1 & 2 NPP. Evaluation of PSA level 2+ (impact on risk for some sequences) performed. Some SM and DiD reviewed based on the latest results for CANDU. Severe Accident Management Guidelines (SAMG) developed and started the systematic review of the technical basis for Emergency Planning (EP). More severe accidents considered to implement with post Fukushima action plan. Evaluations on cliff edge effects and a systematic review of all hazards on going.	M5 ST	Review of the DiD features; post Fukushima actions implemented. All the methods in case U2 used, i.e. PSA level 1, design changes updates compliance with the in force SM and DiD requirements confirmed. PSA level 2 results and some sequences for PSA level 2+ included. SAMG completed and technical basis for EP under review.
		REST	Restart project for Cernavoda U3&4, after construction was stopped in early '90's. SM and DID to be complied with considered to be for a restart project and not for a new project .Impact on SM & DiD new updates under review.	M5 WE	Methodology from L2 U1&2 to be used. Special evaluations on the SM and DiD challenges if not considered a new project may be necessary. Use of extended (to non-technical aspects) of the RIDM.
		REFURB	Refurbishment of Cernavoda NPP U 1 (Pressure Tube replacement and other long-term operation actions implemented).	M5 WE	Methodology from L2 U1, 2 to be used, PSR, and long-term operation plan implemented. SM& DiD challenges to be evaluated with CANDU specific tools based on re-tubing from other plants. Use of RIDM.

Fig. 3.33 Strategies and methods used in the evaluated cases (3)

Example 3 solution $\mathbf{S}_{12}^{\mathbf{PR}_{12}}$ is an example on the role of PSA in providing support for one application, related to the Technical Basis (TB) of the Emergency Plan (EP).

An important element of the TB for EP is the definition of the significant radii (Fig. 3.35) around NPP [21], for the following areas:

(i) PAZ: Precautionary Action Zone,

DETAILED CRITERIA FOR THE EVALUATION OF RESULTS			CONSIDERED CASES													FUTURE EXPECTED TREND
CODE		DESCRIPTION	BAS-U1	EQUIV_ RO	PRA 1	PERIOD I 90 - 00	U2	GIII+ PBMR	AGE	PERIOD II 00 - 11	L2 U1-2	RESTART	REFURB	PERIOD III 11 - 17	PERIOD IV NEXT 10 Y
			C1	C2	C3	CPER1	C4	C5	C6	CPER2	C7	C8	C9	CPER3	CPERN 2030
CR1	CRU	Credibility of uncertainties	1	2	2	2	2	1	1	1	1	1	1	1	1
CR2	CRC	Credibility of the level of conservatism	2	1	1	1	2	1	1	1	2	1	1	1	1
CR3	LEC	Level of conservatism	2	2	1	2	2	2	1	2	1	1	1	1	1
CR4	SM	Safety margin acceptability	2	2	1	2	2	2	1	2	2	2	2	2	2
CR5	CEE	Cliff edge effects	1	1	1	1	1	2	1	1	2	1	1	1	1
CR6	DiDA	Defence in depth - Acceptance criteria for levels and in general	1	1	2	1	2	2	1	2	2	1	1	1	1
CR7	DiDI	Defence in depth - Independence of levels	1	1	2	1	1	2	1	1	2	1	1	1	1
CR8	DPC	The type of method acceptable - deterministic (best estimate or not), probabilistic, combined, using Opex	1	1	2	1	2	2	1	2	2	2	2	2	2
CR9	CHC	Impact of capability to manage change control	1	1	2	1	2	2	1	2	2	2	2	2	2
CR10	CGEN	Inpact of generation / technology phase & HOF	1	1	1	1	2	2	2	2	2	2	2	2	2
CR11	CSIT	Impact of site selection predefined criteria	1	2	2	2	2	2	1	2	2	2	2	2	2
CR12	CEP	Emergency Plan and mitigating actions	1	1	1	1	1	2	1	1	2	2	2	2	1
CRTOT PERIOD			1	1	2	1	2	2	1	2	2	2	2	2	2

Fig. 3.34 Sample case of the safety decisions evolution

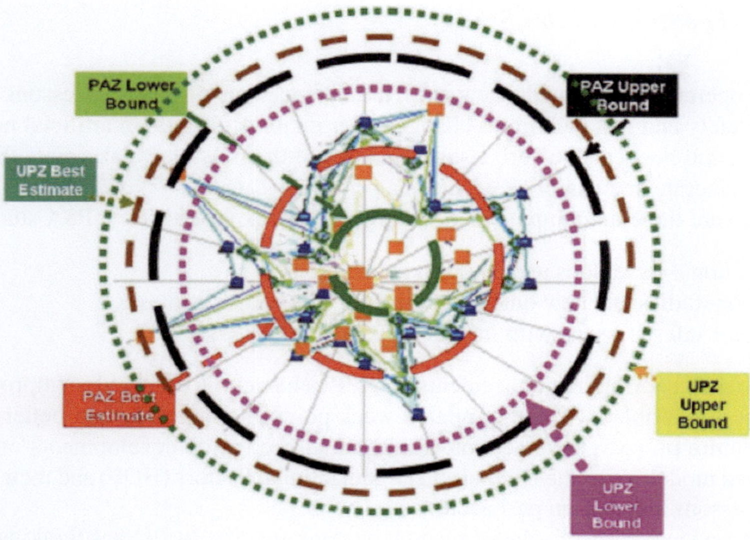

Fig. 3.35 Defining the EP radii by using PSA—sample representation

(ii) UPZ: Urgent Protective action planning Zone,
(iii) LPZ: Long-term Protective Zone (Food Restriction Planning Zone—FRPZ)
(iv) LPZ: Long-term Protective Zone (Food Restriction Planning Zone—FRPZ).

If DSA are used, then the radii have a single well-defined value. However, for PSA levels 2 and 3 approaches, these values have a range of variation. The impact is very important, because it might be such a case, that the lower bound of a radius is higher then the one of the next level on the hierarchy of EP actions (for instance first sheltering then evacuation); as a result it might be a situation that one decision maker.

The results obtained by using PSA show that

(i) The impact of considering the uncertainty in the evaluations of zones radii is important and requires careful attention for the decision-makers. As illustrated in Fig. 3.35, it is possible that upper bound zones for a less restrictive emergency action (like sheltering for instance) could be larger than for the strongest measure (for instance evacuation) if the latter is considered in its lower bound results. Therefore, a very careful sensitivity analysis to define the range of variations of those zones is needed before deciding on their accepted values.
(ii) In the case of multiunit zones, the enveloping radii are different and more conservative in case of using PSA than by using DSA.
(iii) Due to those uncertainties, for the case of NPP close to borders of other countries coordination in adopting radii for those zones is absolutely necessary.

3.1.4 Feedback to the Study

For the operating NPP and the new ones, the enhancement of using the lessons learnt on the safety and risk paradigms changes after major accidents (in artificial nuclear reactors) and how can we derive some features of the possible weak points that are able to generate new major accidents is a priority.

There are three main inputs to the internal feedback process for a PSA study:

- Operation—its lessons and needs in applications,
- Similar studies and new outcomes valid for most other studies,
- General safety lessons with impact on PSA.

The feedback from major accidents on NPP behaviour is expected to improve the forecast of possible NPP safety-related weak points, so that to have a better focus in the future on preventive actions. There are also ongoing developments on new, improved modelling of the Human and Organizational Factors (HOF) and their better use as lessons learnt from past accidents.

PSA review process includes also techniques specific to 'lateral thinking', i.e. possible lessons from other sources than NPP feedback, as, for instance:

- the modelling of complex systems and complex technologies other than nuclear, or
- the operation of artificial reactors and natural reactors (Oklo),

bringing new insights into the input and methodology.

- The Key Topic for the consideration of the Feedback to PSA studies ($\mathbf{KT}_{13}^{FB}$) is to find the proper organizational form to assure a review of PSA study and of the use of its results in applications.
- Problem for the $\mathbf{KT}_{13}^{FB}$ ($\mathbf{PR}\_13^{\mathbf{KT}\_13}$) is that organizing the review for PSA study involves difficulties from technical, staff allocation and financial provisions for any holder of a PSA study.
- Solution for the $\mathbf{PR}\_13^{\mathbf{KT}\_13}$ ($\mathbf{S}\_13^{\mathbf{PR}\_13}$) is to have PSA review included in the Strategic Planning of Safety Analyses Review for the licensing/relicensing process and for its use in the applications at the licensees and/or the nuclear regulators.

Example for solution ($\mathbf{S}\_13^{\mathbf{PR}\_13}$) illustrates organization of the PSA studies in the context of the licensing process for an NPP and the update of its applications, as, for instance:

- Risk monitor of operation and preparation for the maintenance,
- Support to the development and review of the TB for EP,
- Support for the Periodical Safety Review (PSR) as part of the licensing process. The licensees include PSA in their Programs for Strategic Safety Analyses, which involve:
- Allocation of financial support,

- Existence of a core team able to assure review, maintenance and contracting/subcontracting activities,
- Scheduling and support for external review and Peer Review of PSA study and its applications.

3.2 Research Topics in PSA Methodology

The PSA studies development and review are supported by

- the feedback from operation, other studies and the paradigms changes of the general safety evaluations, and
- the continuous research on key aspects of the methodology and inputs. As mentioned in the paragraph on the use of PSA results, during the history of development and use of PSA, the intensive research activities were performed to support PSA methodologies.

Resilience, which is the ability of a system to absorb changes and to maintain its functionality, is probably the most significant post-Fukushima paradigm change affecting the PSA models and use [22].

- This change is related to the need to have in an unitary approach both DBA and BDBA challenges to the NPP.
- This is due to the post-Fukushima conclusion that, after a catastrophic event, it is important for an NPP how fast the recovery will take place.

Resilience has four main facets (Fig. 3.36):

1. Tolerance, which shows how a system behaves near a boundary if challenged.
2. Flexibility, showing the ability of the system to restructure itself when challenged.
3. Margin, describing how closely to the acceptable limits is operating the system.
4. Buffering, as a capacity of the system to absorb without a fundamental breakdown, serious challenges.

To each of these facets, the PSA model has interfaces and it is assumed to support NPP resilience, as illustrated in Fig. 3.36.

This new paradigm makes new priorities in PSA research. The methodologies and the inputs to the PSA studies are *highly impacted by the results from the research activities focused on NPP increasing resilience*.

This paragraph will highlight some *more recent trends in the research* activities, which are considered of interest in the light of this paradigm change.

- The Key Topic for the Research Topics in PSA methodology (**KT**_14^{RSCH}) is to identify in time areas of impact for PSA, able to support inputs to the existing methodology and/or to propose solutions to the new challenges, which appear in PSA in connection mainly with the paradigm changes.

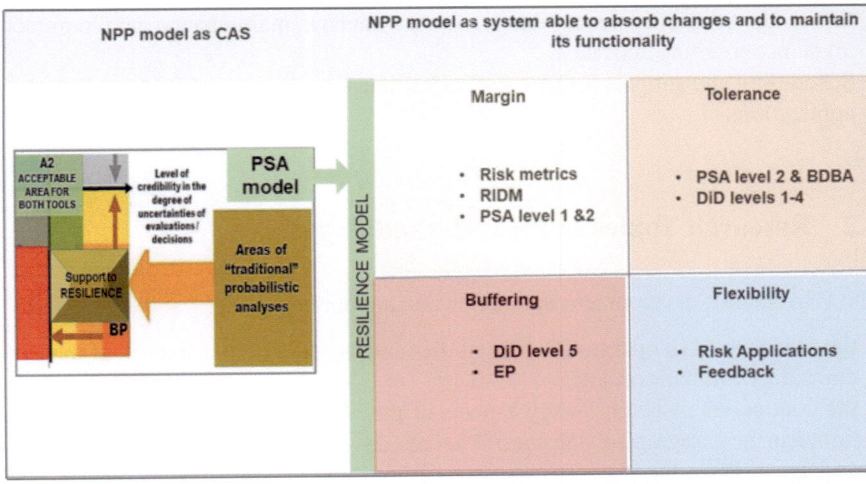

Fig. 3.36 Interface between PSA and resilience models for an NPP

- *Problem for the* $\mathbf{KT}\_14^{RSCH}$ ($\mathbf{PR}\_14^{\mathbf{KT}_{14}}$). The post-Fukushima safety paradigm change led for the PSA to the need to evaluate the CEE and development/adaptation of techniques for the modelling/support of the new generations NPP.
- *Solution for the* $\mathbf{PR}\_14^{\mathbf{KT}_{14}}$ ($\mathbf{S}_{14}^{\mathbf{PR}\_14}$). One possible solution for solving this problem is to explore the tools available for extending the PSA methodology so that:
  - to be able to model not only level 3 of DiD, but also at least levels 2 and 4, and
  - prepare for an integrative approach on DiD levels in the spirit of the existing PSA methods, tools and experience available so far.

Example 1 solution $\mathbf{S}_{14}^{\mathbf{PR}_{14}}$ presents an approach to include models of DiD in the PSA, which is of interest both for the evaluation of the implementation of new requirements for existing NPPs and for the new generation IV of NPPs. The existing situation is that the DiD modelling is supported in a combined DSA–PSA approaches. They are also supported by the results from research and intensive simulations and by OPEX. However, if the results from OPEX do not have answers for issues like DiD modelling, for instance, and/or for new generation NPPs for which there is no OPEX at all, PSA inputs have high modelling and data challenge. The modelling for DiD in a spirit compatible with PSA existing approaches might be illustrated on the IE task. Defining the IE list is of a high challenge for unknown new situations (consideration of CEE) and/or new types of NPPs.

There are already some recommended approaches for generation IV, including SMR to model/consider the DiD aspects.

There are some features of those approaches in relation to the use of PSA and in general of any probabilistic methods, which need to be mentioned, as follows:

- They start from the need to evaluate the compliance of the plants with the Defence in Depth (DiD) principles and all its levels (as illustrated in Fig. 3.36).
- The recommendations for the definition of the details on the DiD as an approach to assure Global Success on Safety (GSS) (formula (3.15)) is considering the impact on safety in the following hierarchy:
  - Safety Functions (SF),
  - Challenges (Ch)
  - Mechanisms ($Mech$),
  - Criteria ($Crit$),
  - Detailed provisions (DP).
- The main objective of the safety approach to assure the GSS is to reach the maximum protection at each level (as shown in Fig. 3.37) [23].
- The building of the DiD following this approach is actually a 'Success Tree' (ST) approach to assure the necessary Levels of Protection (LOPi).

$$GSS = DiD \circ SF \circ Ch \circ Mech \circ Crit \circ DP \tag{3.15}$$

The use of ST is not new in the safety analyses. It was the basis for building scenarios for NPP reaction in some safety philosophies and it was called **Safety Design Matrix** (SDM).

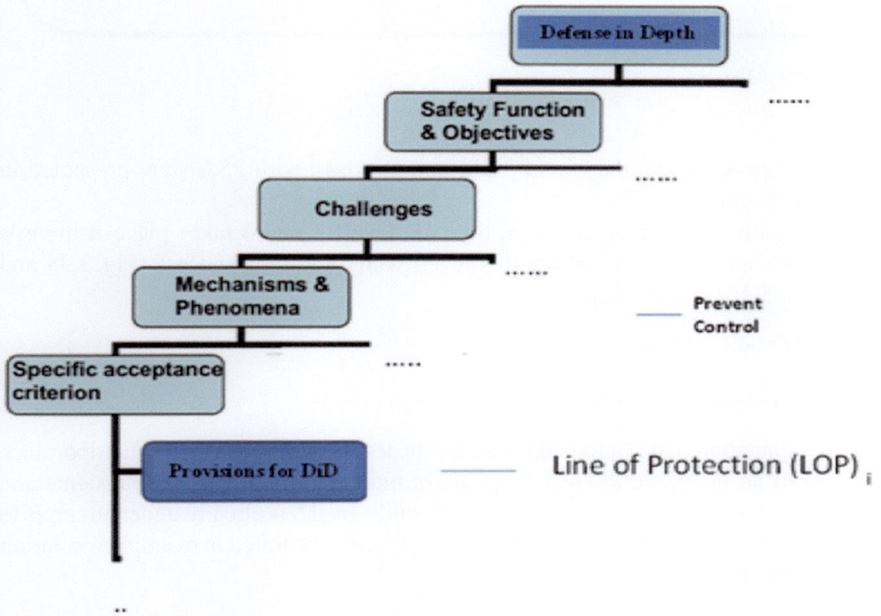

Fig. 3.37 The main criteria used in the process of implementation DiD concept

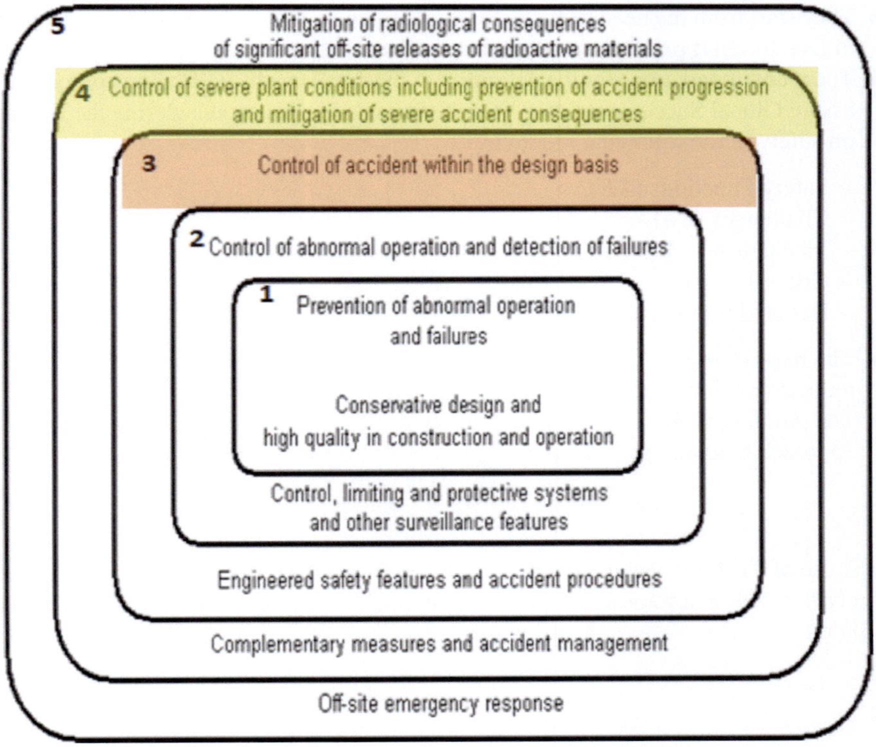

Fig. 3.38 DiD layers

Some references on the use of SDM and interface with PSA were presented in *Example 2 solution* $\mathbf{S}_{12}^{\mathbf{PR}_{12}}$.

Successful risk-free operation at the DiD levels 3 and 4 takes place if there is success for any path challenging this level, which requires success (Fig. 3.38 and formulas (3.16) and (3.17)) [23]:

- for any challenge (IE_i)
  AND
- of the corresponding to it Line of Protection (LOP_i).

A similar approach is adopted for all levels; level 2 is also of very high importance in view of an increased interest to model in more detail the general transients and abnormal states preceeding the DBA cases, while level 5 is already under attempts to have compatible models with PSA levels 2 and 3, as illustrated in example 4 solution $\mathbf{S}_{10}^{\mathbf{PR}_{10}}$ and example 3 solution $\mathbf{S}_{12}^{\mathbf{PR}_{12}}$.

$$SPC_j = \sum_{i=1}^{n} IE_i \cdot DiD_j\_LOP_i - \Delta Uncovered\ by\ DiD_j\_LOP_i \tag{3.16}$$

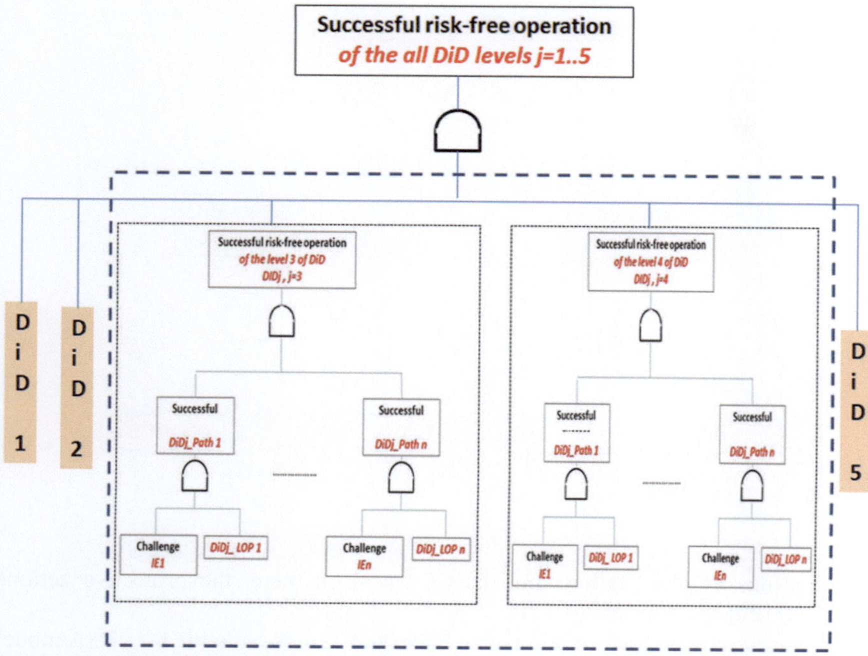

Fig. 3.39 DiD with the layers 3 and 4 presented in detail as Success Trees (ST)

$$\Delta Uncovered\, by\, DiD_j\_LOP_i = Input\, to\, be\, covered\, by\, DiD_{j+1}\_LOP_i \quad (3.17)$$

In order to build a compatible model with the PSA approach, the ST before are transformed (Fig. 3.39).

As a result the

"Successful operation at $DiD3\&DiD4$ **including consideration of how the a level of DiD failed to cover certain** $DiD(J+1)\_LOP$" is substituted in a failure oriented tree (FT in the sense of PSA methodology) by the objective

"Failure to protect (workers, people, environment) for DiD levels 3 and 4" (as in (3.18)).

Due to the fact that a Success-oriented Tree (ST) will require in any case (in a real safety evaluation process) to consider aspects not covered at a certain DiD level, the presence of 'NO' statements makes the 'Failure-oriented trees' (FT) more suitable for the evaluation of ways to identify potential failure paths and protective measures needed.

This objective ($PSA\_OBJ_{PSA_{PATH}}$) is defined as a negation of NON (Successful operation at $DiD3\&DiD4$) and it is actually the Failure to protect at $DiD3\&DiD4$.

Even if the two formulations are equivalent, for the Failure/ Fault Tree (FT) approach, a validated tool might be used, as PSA, in order to build all the possible

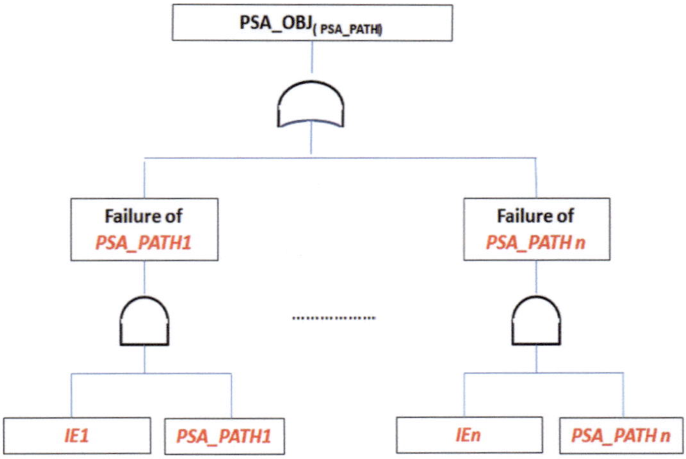

Fig. 3.40 FT for the DiD

combinations of plant failure and derive, based on them, the protective actions (Fig. 3.40 [23] and formula (3.17)).

The result of the transformation of an ST to an FT, more suitable for a PSA model is an expression of the Failure to comply with the objectives in a given DiD level as described in the formula (3.18), where MCSi are results from PSA model.

$$PSA\_OBJ_{PSA_{PATH_n}} = \sum_{i=1}^{n} PSA\_PATH_i = \sum_{i=1}^{n} IE_i \cdot MCS_i \tag{3.18}$$

Example 2 solution $\mathbf{S}_{14}^{\mathbf{PR}_{14}}$ PSA process for a new NPP generation IV type (which is a First of a Kind FOAK installation) is organized in three main steps, as illustrated by the experience of such a development. The process is illustrated in Fig. 3.41.

There are three main steps in developing such types of PSA:

I. Development of PSA model for the initial research phase. At this phase, the inputs mentioned above are used in an initial PSA model. The initial PSA model is actually a Master Fault Tree built to derive all the failure paths. In the resulting set of paths, the IE are identified and considered for all the cases where they appear. The identification of IE is based on the results from previous inputs and on the evaluation of cause–effect relationships and the timing of the phenomena.
II. PSA for the design optimization, with existing IE list and plant model, that is, improved and corrected to reflect the design, while also using the qualitative aspects of its results for the optimization of various design aspects. The experience gained proved that this is a possible process with a high impact on design.
III. PSA for licensing phase, complying with existing and newly developed (but agreed) standards for generation IV NPPs.

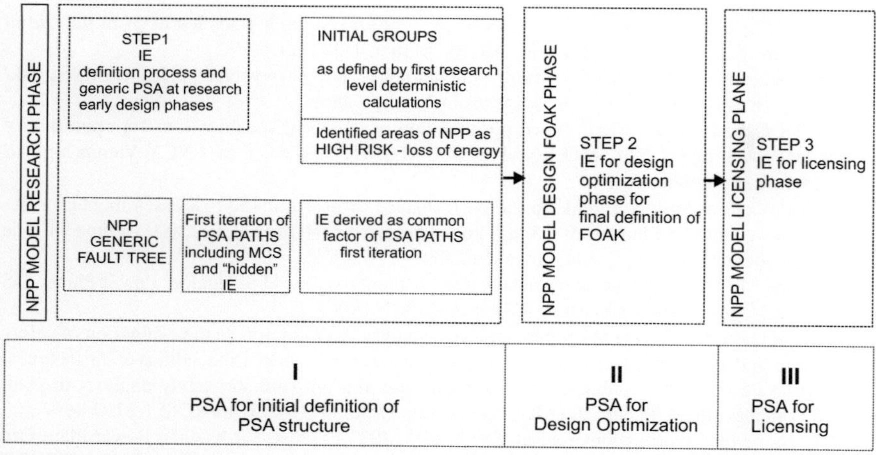

Fig. 3.41 PSA flow path for PSA model for a FOAK NPP

References

1. Serbanescu D (2003) Risk, entropy, synergy and uncertainty in the calculations of gas cooled reactors of PBMR type. https://www2.scopus.com/inward/record.uri?eid=2-s2.0-84933178247&partnerID=40&md5=b9fd8f10427aa074f780b50d6139975b
2. Serbanescu D (2003) Some specifics of the risk analyses for pebble bed modular reactor. In: Programme of the international symposium on nuclear energy SIEN 2003, Nuclear power - a new challenge, Romanian Nuclear Energy Association, AREN, Romania, p 606. http://www.aren.ro/en/programme.pdf
3. Serbanescu D (2015) Selected topics in risk analyses for some energy systems. LAP LAMBERT Academic Publishing
4. Serbanescu D (2005) Some insights on issues related to specifics of the use of probability, risk, uncertainty and logic in PRA studies. Int. J. Crit. Infrastruct. 1(2–3):281–286. https://doi.org/10.1504/IJCIS.2005.006124
5. Serbanescu D (2016) Planificarea pregatirea si raspunsul la urgenta nucleara. Modulul nr. 3 - Procedura stabilirii si utilizarii nivelurilor operationale de urgenta (NOU-zEAL) - Ghid de prezentare schematica a fluxului actiunilor in utilizarea procedurii. https://doi.org/10.13140/RG.2.2.21190.47688
6. Serbanescu D (2017) On some aspects of the multiunit probabilistic safety analyses models. In: 2017 international conference on ENERGY and ENVIRONMENT (CIEM), pp 293–297. https://doi.org/10.1109/CIEM.2017.8120842
7. Serbanescu D (2019) On a possible approach for the multi criteria event analysis in complex systems events. https://doi.org/10.13240/RG.2.2.28999.70403
8. Serbanescu D (2016) A PSA practitioner and safety decision making person view on some issues related to multiple unit PSA analyses. Kick off meeting of the Multiunit PSA project work area 3. In: Vienna IAEA. https://doi.org/10.13140/rg.2.2.32906.06082
9. Nuclear Regulatory Commission DUDoSR Washington (1990) Severe accident risks: an assessment for five US nuclear power plants: appendices A, B, and C. United States. http://inis.iaea.org/search/search.aspx?orig_q=RN:22038232
10. Serbanescu D (2017) Safety paradigm changes and major accidents in nuclear power plants. In: SIEN 2017. https://doi.org/10.13140/RG.2.2.22682.13769

11. Graan HV, Serbanescu D, Eloff L, Combrink Y (2005) Some lessons learnt from the use of PRA during the design phase. Int. J. Crit. Infrastruct. 1(2–3):287–292
12. RiskSpectrum (2019) RiskSpectrum Watcher Doc. http://www.riskspectrum.com/en/risk/Meny_2/RiskSpectrum_DOC/RiskSpectrumDocslide-show
13. TECDOC Series (1993) Risk based optimization of technical specifications for operation of nuclear power plants. 729, INTERNATIONAL ATOMIC ENERGY AGENCY, Vienna. https://www.iaea.org/publications
14. PRA Procedures guide: a guide to the performance of probabilistic risk assessments for nuclear power plants: Chapters 9–13 and appendices A-G (NUREG/CR-2300, volume 2). The American Nuclear Society, LaGrange Park, IL 60525 (1983)
15. NUREG - 1150 : Severe accident risks: an assessment for Five U.S. Nuclear Power Plants. US Nuclear Regulatory Commission, Washington, DC (1990)
16. Some specifics of the use of probabilistic risk analyses as a support to the evaluation of safety margins and the interface with the deterministic based decisions. In: Proceedings of the technical meeting on effective combination of deterministic and probabilistic safety analysis in plant safety management, Paper 29, IAEA (2006). https://doi.org/10.13140/RG.2.1.2794.8647
17. A Phenomena Identification and Ranking Table (PIRT) Exercise for Nuclear Power Plant Fire Modeling Applications (NUREG/CR-6978). US Nuclear Regulatory Commission, USNRC Washington, DC (2008). https://www.nrc.gov/reading-rm/doc-collections/nuregs/contract/cr6978/
18. SOARCA project. US Nuclear Regulatory Commission, USNRC Washington, DC (2019). https://www.nrc.gov/about-nrc/regulatory/research/soar/overview.html
19. Serbanescu Dan (2001) The use of the decision theory and probabilistic analysis in the NPP licensing decision process (IAEA-CN-82/28. Topical Issues in Nuclear Safety, IAEA. https://inis.iaea.org/collection/NCLCollectionStore/_Public/32/046/32046312.pdf
20. Serbanescu D (2015) Risks and human organizational factors (HOF) in nuclear power plants system. https://doi.org/10.13140/RG.2.1.2796.7844
21. Kubanyi J, Lavin RB, Serbanescu D, Toth B, Wilkening H (2008) Risk informed support of decision making in nuclear power plant emergency zoning, generic framework towards harmonising NPP emergency planning practices. DG JRC Institute for Energy
22. Hollnagel E, Woods D, Leveson N (eds) (2006) Resilience engineering: concepts and precepts. http://erikhollnagel.com/books/resilience-engineering-concepts-and-precepts.html
23. Serbanescu D (2017) A specific experience on some challenges in defining an d using defense in depth and safety margin concepts, as highlighted by the safety improvement process. https://doi.org/10.13141/RG.2.1.4859.2488

Chapter 4
Mathematics for Probabilistic Safety Assessments

Abstract The tasks of interest for PSA practitioners are highly based on specialized mathematical tools, which are presented in this chapter. They are related (but not limited) to the following: Presentation of the general theoretical basis for the discrete probability spaces, i.e. formulas, description of the concepts and special aspects related to the random variables and distributions, variance, covariance, correlation and dependent failures, as well as confidence limits. The important aspects of logical structures and how the importance of various contributors to the plant challenges might be calculated are also detailed. The chapter presents also basic definitions and results from special researches on the mathematical background of PSA, as for instance coherent fault trees.

4.1 Basic Probabilities. Discrete Spaces

4.1.1 Basic Definitions and Formulas

1. *Sample Space* Ω: the totality of possible outcomes of a random experiment. The outcomes are called *elementary events*, *basic events*, *points* or *cases* in the classical definition of probability by Laplace.
2. *Discrete sample space* Ω: Ω is at most a denumerable set of points.
3. *Events*: subsets of Ω.
4. *Occurence* of an event A means the appearance of an element/point in A.
5. *Union* of n events $A_1, A_2, \ldots, A_n$ denoted by $A_1 \cup A_2 \cdots \cup A_n$ is the occurence of at least one of the events $A_1, A_2, \ldots, A_n$.
6. *Intersection* of n events $A_1, A_2, \ldots, A_n$ denoted by $A_1 \cap A_2 \cdots \cap A_n$ or $A_1 A_2 \ldots A_n$ is the simultaneous occurence of the events $A_1, A_2, \ldots, A_n$.
7. *Complement* of A, denoted by $!A$ or $\bar{A}$, means that A does not occur.

D. Serbanescu and A. P. Ulmeanu, *Selected Topics in Probabilistic Safety Assessment*, Topics in Safety, Risk, Reliability and Quality 38,
https://doi.org/10.1007/978-3-030-40548-9_4

8. *The difference* of two events A and B is defined by $A - B = A\bar{B}$, i.e. A happens but not B. Consequently, $\bar{A} = \Omega - A$.
9. *Certain event*: The sample space Ω.
10. *Impossible event*: The complement of a certain event, i.e. the empty set Φ.
11. *Probability*: A probability function P is a set function on Ω which satisfies the following three axioms of Kolmogorov:

    i. $\mathbb{P}(\Omega) = 1$;
    ii. For every event A, $\mathbb{P}(A) \geq 0$;
    iii. For every sequence of mutually exclusive events
    $A_1, A_2, \ldots, A_n \quad A_i \cap A_j = \Phi, \quad i \neq j$
    $\mathbb{P}(A_1 \cup A_2 \cup \ldots \cup A_n) = \mathbb{P}(A_1) + \mathbb{P}(A_2) + \cdots \mathbb{P}(A_n)$.

12. *Equally likely elementary events*:
    Let Ω be a finite sample space with N points: $\omega_1, \omega_2, \ldots, \omega_N$.
    If $\mathbb{P}(\omega_i) = 1/N, \quad i = 1, \ldots, N$, then the ω_i are called equally likely elementary events (cases).
13. *Laplace definition of probability*:
    Let $A \subset \Omega$, where the points of Ω are assumed equally likely. Then

$$\mathbb{P}(A) = \frac{Number\ of\ points\ in\ A}{Number\ of\ points\ in\ \Omega}$$

$$\mathbb{P}(A) = \frac{Number\ of\ favorable\ cases\ to\ A}{Number\ of\ cases\ in\ \Omega}$$

14. *Basic formulas for probabilities*:

    i. $\mathbb{P}(\bar{A}) = 1 - \mathbb{P}(A)$.
    ii. $\mathbb{P}(A \cup B) = \mathbb{P}(A) + \mathbb{P}(\bar{A}B) = \mathbb{P}(B) + \mathbb{P}(A\bar{B}) = \mathbb{P}(A) + \mathbb{P}(B) - \mathbb{P}(AB)$.
    iii. $\mathbb{P}(A - B) = \mathbb{P}(A) - \mathbb{P}(AB)$.

15. *Poincaré formulas*:

    i. For any $n > 1$ and for any choice of the events $A_1, \ldots, A_n$,

$$\mathbb{P}(A_1 \cup \cdots \cup A_n) = \sum_{1 \leq i_1 \leq n} \mathbb{P}(A_{i_1}) - \sum_{1 \leq i_1 < i_2 \leq n} \mathbb{P}(A_{i_1} A_{i_2}) + \cdots + (-1)^{n+1} \mathbb{P}\left(\bigcap_{i=1}^{n} A_i\right)$$

    or

$$\mathbb{P}(A_1 \cup \ldots \cup A_n) = X_1 - X_2 + X_3 - \cdots + (-1)^{n+1} X_n$$

    ii. For any $n > 1$ and for any choice of the events $A_1, \ldots, A_n$,

$$\mathbb{P}(A_1 \cdots A_n) = \sum_{1 \leq i_1 \leq n} \mathbb{P}(A_{i_1}) - \sum_{1 \leq i_1 < i_2 \leq n} \mathbb{P}(A_{i_1} \cup A_{i_2}) + \cdots + (-1)^{n+1} \mathbb{P}\left(\bigcup_{i=1}^{n} A_i\right)$$

16. *Waring's* formula:
Let Z_m denote the event that *exactly* m of the events $A_1, \dots, A_n$ occur, for integers $n \geq m \geq 0$ being fixed,

$$\mathbb{P}(Z_m) = \sum_{k=m}^{n} (-1)^{k-m} C_k^m X_k$$

where

$$X_k = \sum_{1 \leq i_1 < i_2 < \cdots < i_k \leq n} \mathbb{P}(A_{i_1} A_{i_2} \cdots A_{i_k})$$

17. *Kounias'* inequalities:

$$\mathbb{P}\left(\bigcup_{i=1}^{n} A_i\right) \leq \sum_{i=1}^{n} \mathbb{P}(A_i) \;-\; \sum_{i=1, 1<j \leq n} \mathbb{P}(A_i A_j)$$

$$\mathbb{P}\left(\bigcup_{i=1}^{n} A_i\right) \leq \sum_{i=1}^{n} \mathbb{P}(A_i) \;-\; \sum_{i=2, j \neq i} \mathbb{P}(A_i A_j)$$

$$\dots$$

$$\mathbb{P}\left(\bigcup_{i=1}^{n} A_i\right) \leq \sum_{i=1}^{n} \mathbb{P}(A_i) \;-\; \sum_{i=n, j \neq n} \mathbb{P}(A_i A_j)$$

Consequently, the following inequality holds:

$$\mathbb{P}\left(\bigcup_{i=1}^{n} A_i\right) \leq min_j \left\{ \sum_{i=1}^{n} \mathbb{P}(A_i) - \sum_{i:i \neq j} \mathbb{P}(A_i A_j) \right\}$$

where $j \in \{1, \dots, n\}$.

Equivalently,

$$\mathbb{P}\left(\bigcup_{i=1}^{n} A_i\right) \leq \sum_{i=1}^{n} \mathbb{P}(A_i) - max_j \sum_{i:i \neq j} \mathbb{P}(A_i A_j)$$

18. *Chung–Erdös* inequality:

$$\mathbb{P}\left(\bigcup_{i=1}^{n} A_i\right) \leq \frac{\left(\sum_{i=1}^{n} \mathbb{P}(A_i)\right)^2}{\sum_{i=1}^{n} \sum_{j=1}^{n} \mathbb{P}(A_i A_j)}$$

19. *Conditional probability* of A given B:

$$\mathbb{P}(A|B) = \frac{\mathbb{P}(AB)}{\mathbb{P}(B)}, \quad \mathbb{P}(B) > 0$$

.

20. *Multiplication formula* for probabilities:

$$\mathbb{P}(AB) = \mathbb{P}(A) \cdot \mathbb{P}(B|A) = \mathbb{P}(B) \cdot \mathbb{P}(A|B)$$

21. *Binomial* identity:

$$(x + y)^n = \sum_{k=0}^{n} C_n^k x^k y^{n-k}$$

22. *Vandermonde's* identity:

$$(x + y)_n = \sum_{k=0}^{n} C_n^k (x)_k (y)_{n-k}$$

where $(x)_k = x(x - 1)(x - 2) \cdots (x - k + 1)$.

23. *Nrlund's* identity:

$$[x + y]_n = \sum_{k=0}^{n} C_n^k [x]_k [y]_{n-k}$$

where $[x]_k = x(x + 1)(x + 2) \cdots (x + k - 1)$.

24. *Multinomial* identity:

$$(x_1 + x_2 + \cdots + x_r)^N = \sum_{n_1+n_2\ldots+n_r=N} \frac{N!}{n_1!n_2! \cdots n_r!} x_1^{n_1} x_2^{n_2} \cdots x_r^{n_r}$$

25. *Pascal's* triangle:

$$C_n^k = C_{n-1}^{k-1} + C_{n-1}^k$$

26. The *Stirling* number of the second kind:
Given a set of N objects, say, the set $\{1, 2, \ldots, N\}$, the Stirling number of the second kind, S_N^n, is defined as the number of all possible partitions of this set into n disjoint and non-empty sets, with no regard of the order in the sets:

i. $S_N^N = S_N^1 = 1$
ii. $S_N^{N-1} = C_N^2$
iii. $S_N^2 = 2^{N-1} - 1$
iv. $S_{N+1}^n = S_N^{n-1} + n \cdot S_N^n$
v. $S_N^n = 1/n! \sum_{k=0}^{n-1} (-1)^k \cdot C_n^k \cdot (n-k)^N$

27. The *Bell* numbers:
The number of partitions of the set A, with $|A| = N$:

$$B_N = \sum_{n=1}^{N} S_N^n$$

28. *Dobinski's* formula for the Bell numbers:

$$B_N = e^{-1} \sum_{m \geq 0} \frac{m^N}{m!}$$

29. *Combinatorial* identity:

$$n^N = \sum_{k=1}^{n} S_N^k \cdot (n)_k$$

where $(n)_k = n(n-1)(n-2)\cdots(n-k+1)$.

4.1.2 Random Variables. Distributions

1. A function *strictly measurable* $X(\omega)$ defined on the sample space $\Omega = \{\omega\}$ is called a random or stochastic variable.
A random variable is called *discrete* if it takes at most a countable set of values, that is, if there exist a sequence $x_1, x_2, \ldots$ such as

$$\mathbb{P}(X = x_k) = p_k > 0$$

and

$$\sum_{k \geq 1} x_k = 1$$

The sequence $\{p_k\}_{k \geq 1}$ defines the probability or the mass function of X.
A random variable X is called absolutely continuous whether for every real c there exist a function f almost everywhere continuous, such that

$$\mathbb{P}(X \leq c) = \int_{-\infty}^{c} f(x)dx$$

The function f is called the probability density of X.

2. The cumulative distribution function of a random variable X is defined by

$$F(x) = \mathbb{P}(X \leq x)$$

For a continuous random variable,

$$F(x) = \int_{-\infty}^{x} f(t)dt$$

If F is differentiable at x, then

$$\frac{dF(x)}{dx} = f(x)$$

It follows that the probability $f(x)dx$ can be interpreted as

$$f(x)dx = \mathbb{P}(x < X \leq x + dx)$$

For a discrete random variable,

$$F(x) = \sum_{x_i \leq x} \mathbb{P}(X = x_i)$$

3. The main discrete distributions

   i. the *binomial* distribution with parameters n (number of independents trials) and p (probability of success on each trial): $X \sim Bin(n, p)$

   $$\mathbb{P}(X = k) = C_n^k p^k (1 - p)^{n-k} \qquad k = 0, 1, \ldots, n$$

   ii. the *multinomial* distribution for the random vector $\mathbf{X} = (X_1, X_2, \ldots, X_n)$ with integer parameter $N > 0$ and probability vector $\mathbf{p} = (p_1, p_2, \ldots, p_n)$: $\mathbf{X} \sim Mult(N, \mathbf{p})$

   $$\mathbb{P}(X_1 = x_1, X_2 = x_2, \ldots, X_n = x_n) = \frac{N!}{x_1!x_2! \cdots x_n!} p_1^{x_1} p_2^{x_2} \cdots p_n^{x_n}$$

   with $x_1, x_2, \ldots x_n$ positive integers, following $\sum_{i=1}^{n} x_i = N$ and the probabilities $p_1 + p_2 + \cdots + p_n = 1$.

   Remark:
   The individual components of a multinomial random vector $\mathbf{X}$ are binomial and have a binomial distribution: $X_1 \sim Bin(N, p_1)$, $X_2 \sim Bin(N, p_2)$, ... $X_n \sim Bin(N, p_n)$.

   iii. the *geometric* distribution with parameter p (probability of success on each trial): $X \sim Geo(p)$

   $$\mathbb{P}(X = k) = (1 - p)^{k-1} p \qquad k = 1, 2, \ldots$$

iv. the *Pascal* distribution with parameters p (probability of succes on each trial) and n (positive integer parameter): $X \sim Pascal(n, p)$:

$$\mathbb{P}(X = k) = C_{k-1}^{n-1} p^n (1-p)^{k-n} \qquad k = n, n+1, \ldots$$

The Pascal distribution can be used to model the total number of trials k before the nth success, in repeated mutually independent Bernoulli trials, each with probability of success p.

v. the *Poisson* distribution with parameter $\lambda > 0$: $X \sim Po(\lambda)$:

$$\mathbb{P}(X = k) = \lambda^k / k! e^{-\lambda} \qquad k = 0, 1, 2, \ldots$$

vi. the discrete *Weibull* distribution with real parameter p $(0 < p < 1)$, and positive shape parameter β: $X \sim Discrete\ Weibull(p, \beta)$

$$\mathbb{P}(X = k) = p^{k^\beta} - p^{(k+1)^\beta} \qquad k = 0, 1, \ldots$$

The discrete Weibull distribution is a flexible model of count data that can handle both over- and under-dispersion.

4. The main continuous distributions

i. the *exponential* distribution with one parameter $\lambda > 0$: $X \sim Exp(\lambda)$

$$f(x) = \lambda e^{-\lambda x} \qquad x \geq 0$$

ii. the *exponential* distribution with two parameters $\theta > 0$ and $\eta \in \mathbb{R}$: $X \sim Exp(\theta, \eta)$

$$f(x) = \frac{1}{\theta} e^{-\frac{x-\eta}{\theta}} \qquad x \geq \eta$$

iii. the *Weibull* distribution with shape parameter $\beta > 0$ and scale parameter $\alpha > 0$: $X \sim Weibull(\alpha, \beta)$

$$f(x) = \frac{\beta}{\alpha} \cdot \left(\frac{x}{\alpha}\right)^{\beta-1} e^{-(x/\alpha)^\beta} \qquad x \geq 0$$

Remarks:
– if X follows the standard exponential distribution (parameter $\lambda = 1$), then $Y = \alpha \cdot X^{1/\beta}$ follows a Weibull distribution with shape parameter β and scale parameter α;
– if Y follows the Weibull distribution with shape parameter β and scale parameter α, then $X = (Y/\alpha)^\beta$ follows the standard exponential distribution $X \sim Exp(\lambda = 1)$.

iv. the *Gamma* distribution with shape parameter $\alpha > 0$ and scale parameter $\lambda > 0$: $X \sim Gamma(\alpha, \lambda)$:

$$f(x) = \frac{\lambda^\alpha}{\Gamma(\alpha)} x^{\alpha-1} e^{-\lambda x} \qquad x \geq 0$$

where the Gamma function $\Gamma(\alpha) = \int_0^\infty x^{\alpha-1} e^{-x} dx$.

Remarks:
– the exponential distribution with parameter λ is identical to the Gamma distribution with parameters $(1, \lambda)$;
– if $X_1, \ldots, X_n$ are independent exponential random variables, each with parameter λ, then the sum $X = X_1 + \cdots + X_n$ is a random variable following a Gamma distribution with parameters (n, λ).

v. the *Gaussian* distribution with standard deviation $\sigma > 0$ and expectation $\mu \in \mathbb{R}$: $X \sim N(\mu, \sigma)$

$$f(x) = \frac{1}{\sigma\sqrt{2\pi}} e^{-\frac{1}{2}\left(\frac{x-\mu}{\sigma}\right)^2} \qquad x \in \mathbb{R}$$

Let Φ denote the standard normal distribution $\mu = 0$ and $\sigma = 1$. Then, the normal cumulative distribution function F is given by

$$F(x) = \Phi\left(\frac{x-\mu}{\sigma}\right) \qquad x \in \mathbb{R}$$

vi. the *log − normal* distribution with parameters $\mu \in \mathbb{R}$ and $\sigma > 0 : X \sim LogN(\mu, \sigma)$

$$f(x) = \frac{1}{\sigma x\sqrt{2\pi}} e^{-\frac{1}{2}\left(\frac{ln(x)-\mu}{\sigma}\right)^2} \qquad x > 0$$

Remarks:
– The probabilistic safety studies extensively use the log-normal distribution to represent the uncertainty in the estimation of failure probabilities. Morever, as a consequence of the Central Limit Theorem, the logical multiplication of a large number of components having arbitrary but well-behaved lifetime distributions results in a log-normal distribution;
– Useful percentiles of the log-normal distribution and the error factor formula are given in the Table 4.1;

– if Y follows a normal distribution with parameters $\mu \in \mathbb{R}$ and $\sigma > 0$, then $X = e^Y$ follows the log-normal distribution with mean μ and standard deviation σ;
– the log-normal cumulative distribution function F is given by

$$F(x) = \Phi\left(\frac{ln(x)-\mu}{\sigma}\right) \qquad x > 0$$

Table 4.1 Useful percentiles of the log-normal distribution and the error factor formula

Percentile	Value
5th	$x_5 = exp(\mu - 1.645\sigma) = x_{50}/EF$
50th	$x_{50} = exp(\mu) = \sqrt{x_5 \cdot x_{95}}$
95th	$x_{95} = exp(\mu + 1.645\sigma) = x_{50} \cdot EF$
Error factor	$EF = \sqrt{x_{95}/x_5}$

vii. the distribution Beta with shape parameters $\alpha > 0$ and $\beta > 0$: $X \sim Beta(\alpha, \beta)$

$$f(x) = \frac{1}{B(\alpha, \beta)} x^{\alpha-1} \cdot (1-x)^{\beta-1} \qquad x \in [0, 1]$$

where $B(\alpha, \beta)$ is the Beta function: $B(\alpha, \beta) = \int_0^1 t^{\alpha-1}(1-t)^{\beta-1} dt$.

Figure 4.1 provides an illustration of the Mathematica calculus for the percentiles x_5, x_{50}, x_{95} and the error factor ERF.

Remarks:
– the Beta distribution is reduced to the continuous uniform distribution when $\alpha = \beta = 1$;
– if X_1 and X_2 are independent gamma-distributed random variables with parameters (a, θ) and (b, θ), respectively, then the random variable $X = X_1/(X_1 + X_2)$ is Beta-distributed with parameters (a, b).

5. The main compound distributions

   i. The *Beta-binomial* distribution is a compound distribution of the Beta and the binomial distributions. It is a natural extension of the binomial model. It is obtained when the parameter p in the binomial distribution is assumed

```
Clear["Global`*"]
α = 2;
β = 90;
X = Quantile[BetaDistribution[α, β], #] & /@ {0.05, 0.5, 0.95}
{0.00391901, 0.0183748, 0.0510702}
ERF = √X[[3]] / X[[1]]
3.6099
```

Fig. 4.1 An illustration of the mathematica calculus for the percentiles x_5, x_{50}, x_{95} and error factor ERF, in the case of the Beta distribution

to be a random variable, denoted by $\mathbb{P}$, that follows a Beta distribution with parameters α and β, i.e. $\mathbb{P} \sim Beta(\alpha, \beta)$: $X \sim BB(\alpha, \beta, n)$.
For n independent trials,

$$\mathbb{P}(X = k|\alpha, \beta, n) = \int_0^1 \mathbb{P}(X = k|\mathbb{P} = p) f_{\mathbb{P}}(p) dp$$

$$\mathbb{P}(X = k|\alpha, \beta, n) = \int_0^1 C_n^k p^k (1-p)^{n-k} \frac{1}{B(\alpha, \beta)} p^{\alpha-1} (1-p)^{\beta-1} dp$$

It follows that

$$\mathbb{P}(X = k|\alpha, \beta, n) = C_n^k \frac{B(k+\alpha, n-k+\beta)}{B(\alpha, \beta)} \qquad k = 0, 1, 2, \ldots, n$$

ii. The *Beta-geometric* distribution is a compound distribution of the Beta and the geometric distributions. It is a natural extension of the geometrical model. It is obtained when the parameter p in the geometric distribution is assumed to be a random variable, denoted by $\mathbb{P}$, that follows a Beta distribution with parameters α and β, i.e. $\mathbb{P} \sim Beta(\alpha, \beta)$: $X \sim BG(\alpha, \beta)$.

$$\mathbb{P}(X = k|\alpha, \beta) = \int_0^1 \mathbb{P}(X = k|\mathbb{P} = p) f_{\mathbb{P}}(p) dp$$

$$\mathbb{P}(X = k|\alpha, \beta) = \int_0^1 (1-p)^{k-1} p \frac{1}{B(\alpha, \beta)} p^{\alpha-1} (1-p)^{\beta-1} dp$$

It follows that

$$\mathbb{P}(X = k|\alpha, \beta) = \frac{B(\alpha+1, \beta+k)}{B(\alpha, \beta)} \qquad k = 0, 1, 2 \ldots$$

4.1.3 Expectation. Variance

1. The expected value or the mean value of a random variable X, denoted by $\mathbb{E}(X)$, is defined by

$$\mathbb{E}(X) = \begin{cases} \sum_i x_i \, \mathbb{P}(X = x_i) & for\ a\ discrete\ variable\ X \\ \int_{\mathbb{R}} x f(x) dx & for\ a\ continuous\ variable\ X \end{cases}$$

Remarks:
– the linearity property of the expectation operation: if $\mathbb{E}(X) < \infty$, $\mathbb{E}(Y) < \infty$, then for any constants a, b we have

$$\mathbb{E}(aX + bY) = a\,\mathbb{E}(X) + b\,\mathbb{E}(Y)$$

– if g is a measurable function and X is a random variable X, then we have the mean value of a function g(X):

$$\mathbb{E}(Y) = \mathbb{E}(g(X)) = \begin{cases} \sum_i g(x_i) f(x_i) & for\ X\ discrete\ variable\ X \\ \int_{\mathbb{R}} g(x) f(x) dx & for\ X\ continuous\ variable\ X \end{cases}$$

– for a continuous variable X with density function f and cumulative distribution function F, the mean value of X is given by

$$\mathbb{E}(X) = \int_{-\infty}^{\infty} x f(x) dx = \int_{-\infty}^{0} x dF(x) - \int_{0}^{\infty} x d(1 - F(x))$$

Integrating by parts, since $\lim_{x \to -\infty} xF(x) = lim_{x \to +\infty} x(1 - F(x)) = 0$, it follows that

$$\mathbb{E}(X) = \int_{0}^{\infty} (1 - F(x)) dx - \int_{-\infty}^{0} F(x) dx$$

Consequently, as shown the Fig. 4.2, the mean value is geometrically interpreted as the difference between the two areas: $\mathbb{E}(X) = A - B$.

2. The variance of the random variable X, denoted by $Var(X)$, measures the spread or variability of its distribution, and is defined by

$$Var(X) = \begin{cases} \sum_i (x_i - \mathbb{E}(X))^2 p_i & for\ X\ discrete\ variable\ X \\ \int_{\mathbb{R}} (x - \mathbb{E}(X))^2 f(x) dx & for\ X\ continuous\ variable\ X \end{cases}$$

The standard deviation $\sigma(X)$ is the square root of the variance.

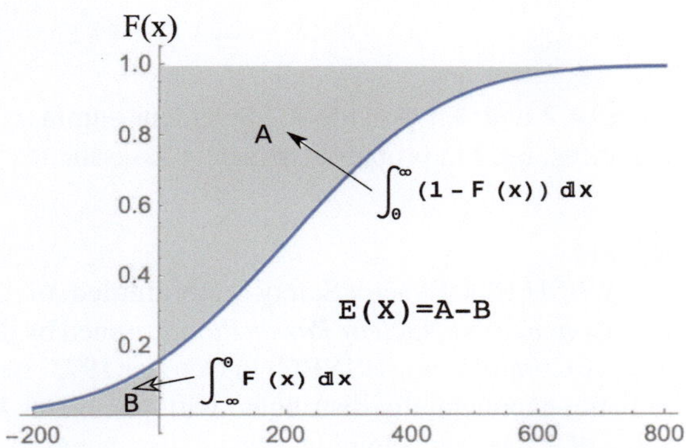

Fig. 4.2 The geometrical interpretation of the mean value

Table 4.2 Mean and variance for several discrete distributions

Distribution	$\mathbb{P}(X = k)_{k \in \mathcal{K}}$	$\mathcal{K}$	$\mathbb{E}(X)$	$Var(X)$
Binomial $Bin(n, p)$	$C_n^k p^k (1-p)^{n-k}$	$k \in \{0, 1, \ldots, n\}$	$n \cdot p$	$n \cdot p \cdot (1-p)$
Poisson $Po(\lambda)$	$\lambda^k / k! \cdot e^{-\lambda}$	$k \in \{0, 1, \ldots\}$	λ	λ
Geometric $Geo(p)$	$(1-p)^{k-1} p$	$k \in \{1, 2, \ldots\}$	$p/(1-p)$	$p/(1-p)^2$
Pascal $Pascal(n, p)$	$C_{k-1}^{n-1} p^n (1-p)^{k-n}$	$k \in \{n, n+1, \ldots\}$	n/p	$n(1-p)/p^2$

Table 4.3 Mean and variance for several continuous distributions

Distribution	$f(x)_{x \in \mathcal{X}}$	$\mathcal{X}$	$\mathbb{E}(X)$	$Var(X)$
$Exp(\lambda)$	$\lambda e^{-\lambda x}$	$[0, \infty)$	$1/\lambda$	$1/\lambda^2$
$Gamma(\alpha, \lambda)$	$\lambda^\alpha / \Gamma(\alpha) x^{\alpha-1} e^{-\lambda x}$	$[0, \infty)$	α/λ	α/λ^2
$Weibull(\alpha, \beta)$	$\beta/\alpha \cdot (x/\alpha)^{\beta-1} \cdot exp(-(x/\alpha)^\beta)$	$[0, \infty)$	$\alpha \cdot \Gamma(1 + 1/\beta)$	$\alpha^2(\Gamma(1+2/\beta) - \Gamma^2(1+1/\beta))$
$Gauss(\mu, \sigma)$	$1/\sqrt{2\pi} \cdot exp(-(x - \mu)^2/2/\sigma^2)$	$\mathbb{R}$	μ	σ^2
$LogN(\mu, \sigma)$	$1/(x\sqrt{2\pi}) \cdot exp(-(Ln(x) - \mu)^2/2/\sigma^2)$	$(0, \infty)$	$e^\mu e^{\sigma^2/2}$	$e^{2\mu} e^{2\sigma^2} - e^{2\mu} e^{\sigma^2}$
$Beta(\alpha, \beta)$	$x^{\alpha-1}(1-x)^{\beta-1} / B(\alpha, \beta)$	$[0, 1]$	$\alpha/(\alpha+\beta)$	$\alpha \cdot \beta/(\alpha+\beta)^2 / (1+\alpha+\beta)$

Chebyshev's inequality
– Let X be a continuous random variable with finite expected value $\mathbb{E}(X)$ and finite variance $Var(X)$. Then, for any real number $\epsilon > 0$,

$$P(|X - \mathbb{E}(X)| \geq \epsilon) \leq \frac{Var(X)}{\epsilon^2}$$

Tables 4.2 and 4.3 provide a convenient summary of distributions, means and variances, used in probabilistic safety assessment.

Remark:
The WASH 1400 Reactor Safety Study entitled *An Assessment of Accident Risks in U.S. Commercial Nuclear Power Plants*, issued by the United States Nuclear Regulatory Commission (USNRC) in October 1975, treated the probability of failure as being exponentially distributed with parameter λ time-invariant. It treated the value of λ itself as being log-normal distributed.

4.1.4 Confidence Limits

In the context of process industries, such as oil and gas, but as well in the nuclear, chemical and aeronautical fields, complex automated safety functions are applied to achieve hazard risk reduction. The functional safety standards place a strong emphasis on the need to obtain credible failure rate data for use in probabilistic safety assessments.

Over the past decades, an important amount of information has been collected in the above-mentioned fields to enable failure rates to be estimated for all of the commonly used components in safety functions. The information shows the failure rates that are being achieved in practice. It also shows that the failure rates measured for any particular type of device vary by at least an order of magnitude. The variation depends largely on the service, operating environment and maintenance practices.

The failure rates from industry databases are useful in demonstrating the feasibility of the risk reduction being targeted by safety functions, which is important in setting operational reliability benchmarks.

The failure rates measured from a facility's maintenance data are useful in demonstrating the risk reduction that a safety function can achieve, for a given operating service, environment and set of maintenance practices.

The basic purpose of functional safety is to provide defined levels of risk reduction for the hazards associated with the nuclear power plants.

Functional safety usually relies on systems of electrical, electronic or programmable functions and interlocks. These systems can be complicated and subject to hidden or latent failures. Functional safety maintains safety integrity of assets in two ways:

- Systematic safety integrity deals with preventable failures. These are failures resulting from errors and shortcomings in the design, manufacture, installation, operation, maintenance and modification of the safety systems;
- Hardware safety integrity deals with controlling random hardware failures. These are the failures that occur at a reasonably constant rate and are completely independent of each other. They are not preventable and cannot be avoided or eliminated, but the probability of these failures occurring can be calculated.

Consequently, the functional safety relies on a concrete demonstration that the automated safety systems can reliably achieve the specified risk reduction. The order of magnitude of Risk Reduction Factor (RRF) determines the Safety Integrity Levels (SIL) of a safety function, as shown in Table 4.4.

The risk reduction factor is inversely proportional to the Probability of Failure on Demand (PFD). A safety function with a probability of failure on demand of 0.01 achieves a RRF of 100. State-of-the-art methods for reliability calculations are described in more detail in the Technical Report ISO 12489 'Petroleum, petrochemical and natural gas industries—reliability modelling and calculation of safety systems' and IEC 61508-6:2010 [1].

Several other useful references are available on this subject, including ISA-TR84.00.02-2015 *Safety Integrity Level (SIL) Verification of Safety Instrumented*

Table 4.4 The safety integrity levels of a safety function

Risk reductor factor	Safety integrity level
RRF range 10 to 100	SIL 1
RRF range 100 to 1000	SIL 2
RRF range 1000 to 10000	SIL 3
RRF range 10000 to 100000	SIL 4

Functions [2] and SINTEF 2013 Reliability Prediction Method for Safety Instrumented Systems—PDS Method Handbook [3].

Confidence limits are partial integrations over a probability density function. There are two cases: failure on demand and failure with time (unreliable).

In actual PSA practice in the nuclear field, it is often the case that the Beta distribution is applied in a straightforward manner in order to estimate the probability of failure on demand. The following exemple illustrates the application of the method. We denote: n—the number of demands; k—the number of failures, $0 \leq k \leq n; data = \{k_1/n_1, k_2/n_2, k_3/n_3, \ldots, k_N/n_N\}$—the record of data concerning the unavailability of a such system in operation in N similar nuclear power plants:

- estimate α and β parameters for Beta distribution, as shown in Fig. 4.3.
- find the 90% confidence interval $[PFD_{5\%}, PFD_{95\%}]$, as shown in Fig. 4.4.

For the BWRs listed in Table 4.5, the PSA results are expressed in Fig. 4.5, in terms of 90% confidence interval for HPCI unavailability, following the statistical treatment of the recorded data concerning relevant HPCI failure modes: failure of the injection valve to open; failure to start due to components other than the injection valve; failure of the turbine drive pump to run given it started and system out of service due to testing/maintenance.

```
data = {3 / 24, 2 / 48, 1 / 10, 3 / 19, 2 / 45};

params = FindDistributionParameters[data, BetaDistribution[α, β]]

{α → 3.5047, β → 33.9083}
```

Fig. 4.3 An illustration of the Mathematica code to estimate the Beta distribution parameters

```
PFD1 = Quantile[BetaDistribution[α, β] /. params, 0.05]

0.0304361

PFD2 = Quantile[BetaDistribution[α, β] /. params, 0.95]

0.181622
```

Fig. 4.4 An illustration of the Mathematica code to find the 90% confidence interval for the Probability of Failure on Demand (PFD)

Table 4.5 Beta distribution parameters for comparing HPCI system unavailability for nine US commerical BWRs

No.	Plant	α	β
1	Browns Ferry 2	3.46	48.93
2	Brunswick 1	1.93	7.55
3	Brunswick 2	2.16	11.28
4	Cooper	2.99	29.95
5	Fermi 2	3.54	27.33
6	FitzPatrick	4.14	66.72
7	Hatch	12.27	139.43
8	Peach Bottom 2	1.43	11.55
9	Vermont Yankee	8.73	106.41

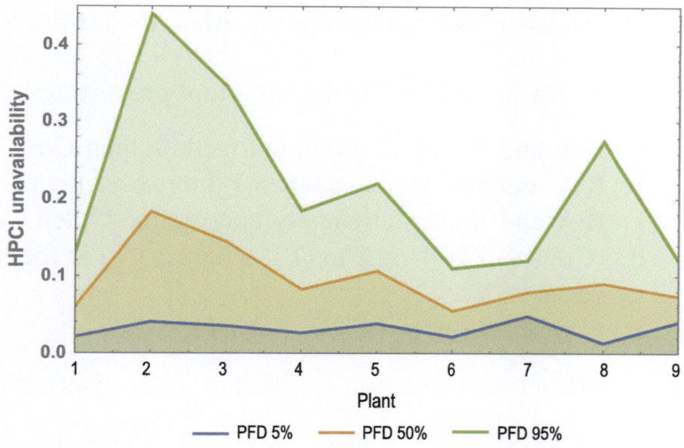

Fig. 4.5 90% confidence intervals for HPCI system unavailability for nine US commercial BWRs (presented in Table 4.5)

4.1.5 Covariance. Correlation

Generally speaking, the covariance $Cov(A, B)$ between two features A and B measures their tendency to vary together, i.e. to co-vary. Where the variance is the average of the squared deviation of the feature from its mean, the covariance is the average of the products of deviations of features from their means.

1. In the case of two real random variables X and Y, we have

$$Cov(X, Y) = \mathbb{E}((X - \mathbb{E}(X))(Y - \mathbb{E}(Y)))$$

   If $Var(X)$ and $Var(Y)$ are finite, then

$$Cov(X, Y) = \mathbb{E}(X \cdot Y) - \mathbb{E}(X) \cdot \mathbb{E}(Y)$$

$$Var(X + Y) = Var(X) + Var(Y) + 2Cov(X, Y)$$

For any real constants $c_1, c_2, \dots c_n$ and real random variables $X_1, X_2, \dots, X_n$ with finite $Var(X_i)$ $(i = 1, \dots, n)$,

$$Var\left(\sum_{i=1}^{n} c_i X_i\right) = \sum_{i=1}^{n} c_i^2 Var(X_i) + 2\sum_{i=1}^{n-1}\sum_{j>i}^{n} c_i c_j Cov(X_i, X_j)$$

The covariance coefficient is defined as $\sigma_{ij} = Cov(X_i, X_j)$. For $i = j$, it follows that $\sigma_{ii} = Cov(X_i, X_i) = Var(X_i)$.

With the covariance coefficients, for a random vector $X_1, X_2, \dots X_n$, we can calculate the entries of the covariance matrix $C_{i,j=1,\dots,n}$, which is a square $n \times n$ matrix given by $C_{i,j} = \sigma_{ij}$.
The diagonal entries of the covariance matrix are the variances, the other entries are the covariances. For this reason, the covariance matrix is sometimes called the variance–covariance matrix. Also, the covariance matrix is symmetric since $C_{ij} = C_{ji}$.
The covariance has several important properties:

a. If X and Y tend to increase together, then $Cov(X, Y) > 0$;
b. If X tends to decrease when Y increases, then $Cov(X, Y) < 0$;
c. If X and Y are statistically uncorrelated, then $Cov(X, Y) = 0$;
d. $|Cov(X, Y)| \leq \sigma(X)\sigma(Y)$, where $\sigma(X)$ is the standard deviation of random variable X;
e. $Cov(X, X) = \sigma^2(X) = Var(X)$.

2. The Pearson correlation coefficient $\rho(X, Y)$ is defined as

$$\rho(X, Y) = \frac{Cov(X, Y)}{\sqrt{Var(X)}\sqrt{Var(Y)}}$$

for two real random variables X, Y with finite variances.

It is worth to mention that the correlation does not imply causation. For instance, Fig. 4.6 shows a high coefficient correlation between two random and completely unrelated features.

3. The concept of information entropy has been introduced by Claude Shannon. His concept describes how much information is there in a signal or in a sequence of events. Shannon defines entropy in terms of discrete random variable X, with possible states/outcomes $x_1, x_2, \dots, x_n$:

$$H(X) = \sum_{i=1}^{n} p(i) \cdot log_2\,(1/p(i)) = -\sum_{i=1}^{n} p(i) \cdot log_2 p(i)$$

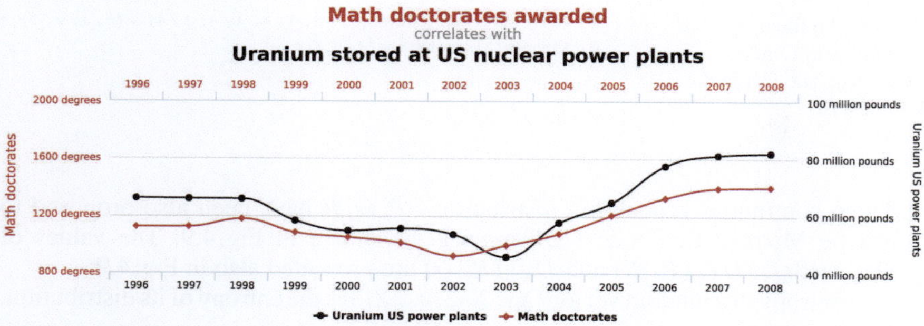

Fig. 4.6 An illustration of a high correlation between two random and completely unrelated features. (data sources: USA National Science Foundation and Department of Energy)

where $p(i) = \mathbb{P}(X = x_i)$ is the probability of the ith outcome of X, with the convention $0 \cdot log 0 = 0$.

4. The Kullback–Leibler (KL) divergence is the expectation of the log difference between the original distribution $\mathcal{P}$ relative to another distribution $\mathcal{Q}$.

$$D_{KL}(P||Q) = \sum_{i=1}^{n} q(i) \cdot (log_2 q(i) - log_2 p(i)) = \sum_{i=1}^{n} q(i) \cdot log_2 \frac{q(i)}{p(i)}$$

where $q(i) = \mathbb{P}(X = x_i)\ with\ X \sim \mathcal{Q}$ and $p(i) = \mathbb{P}(X = x_i)\ with\ X \sim \mathcal{P}$. In the continuous case, the Kullback–Leibler divergence between the original distribution $\mathcal{P}$ relative to another distribution $\mathcal{Q}$ is defined as

$$D_{KL}(P||Q) = \int_{\mathbb{R}} g(x) \cdot log_2 \frac{g(x)}{f(x)} dx$$

where $f(x)$ and $g(x)$ are the probability density functions:
– in the case $X \sim \mathcal{P}$: $f(x)dx = \mathbb{P}(x < X \leq x + dx)$;
– in the case $X \sim \mathcal{Q}$: $g(x)dx = \mathbb{P}(x < X \leq x + dx)$.
The implementation of the Kullback–Leibler divergence using the Mathematica's probability and distribution functions is presented in Fig. 4.7.
Two discrete probability distributions (uniform $\mathcal{P}$ and binomial $\mathcal{Q}$) have been proposed to test the Mathematica code *klDivergence*, as shown in Fig. 4.8. The value of $D_{KL}(P||Q)$ is presented also in Fig. 4.8.

Fig. 4.7 The source code in Mathematica for a function named *klDivergence* that follows the definition of the Kullback–Leibler divergence

```
Options[klDivergence] = Options@Expectation;
klDivergence[distP_, distQ_,
  opts : OptionsPattern[klDivergence]] :=
 Expectation[Log@PDF[distQ, x] / Log[2] -
   Log@PDF[distP, x] / Log[2], x ≈ distQ, opts]
```

```
P = EmpiricalDistribution[{1/4, 1/4, 1/4, 1/4} → {0, 1, 2, 3}];
Q = BinomialDistribution[3, 0.5];

klDivergence[P, Q] (* the result is given in bits *)

0.188722
```

Fig. 4.8 An illustration of the Kullback–Leibler divergence calculus in the discrete case

Three continuous probability distributions $\mathcal{P}, \mathcal{Q}, \mathcal{R}$ have been also proposed to test he Mathematica code *klDivergence*, as shown in Fig. 4.9. The values of $D_{KL}(P||Q)$, $D_{KL}(P||R)$ and $D_{KL}(R||Q)$ are presented also in Fig. 4.9.

The entropy of a random vector $(X_1, X_2, \ldots, X_n)$ is the entropy of its distribution, that is

$$H(X_1, \ldots, X_n) = -\sum_{E^n} \mathbb{P}(X_1 = x_1, \ldots, X_n = x_n) \cdot log_2 \mathbb{P}(X_1 = x_1, \ldots, X_n = x_n)$$

where if X_i takes values in a discrete set $E \subset \mathbb{R}$ for all $i = 1, \ldots, n$, then the sum is taken on all $(x_1, \ldots, x_n) \in E^n$, and

$$H(X_1, \ldots, X_n) = -\int_I f(x_1, \ldots, x_n) \cdot log_2 f(x_1, \ldots, x_n) dx_1 \ldots dx_n$$

if $(X_1, \ldots, X_n)$ has a density f, positive on $I \subset \mathbb{R}^n$.

Remark:

For a finite space, the entropy is maximum for uniform probability. Furthermore, the entropy is increasing with n, this means that the uncertainty of a system increases with the number of its components.

Example: Let us consider a system with $n = 4$ components and the random vector $\mathbf{X} = (X_1, \ldots, X_4)$ that follows a multinomial distribution with uniform

```
distP = BetaDistribution[0.5, 0.5];
distQ = UniformDistribution[];
distR = TriangularDistribution[{0, 1}];

klDivergence[distP, distQ]

0.208801

klDivergence[distP, distR]

0.766106

klDivergence[distR, distQ]

 2 (-1 + Log[2])
-───────────────
     Log[4]
```

Fig. 4.9 An illustration of the Kullback–Leibler divergence calculus in the continuous case

Fig. 4.10 The illustration of the Mathematica calculus for the Shannon entropy, in the case of the system with n = 4 components and uniform probabilities

```
n = 4; p = 1 / n;

d = MultinomialDistribution[n, {p, p, p, p}];

x =.;

H = 0;

For[i = 0, i ≤ 4, i++,
 For[j = 0, j ≤ 4, j++,
  For[k = 0, k ≤ 4, k++,
   For[l = 0, l ≤ 4, l++,
    If[i + j + k + l == 4,
     x = PDF[d, {i, j, k, l}];
     H += -x Log[2, x ]]]]]]

N[H]

4.81511
```

probability $p = 1/n$, $Mult\,(n, \{p_1, \ldots, p_4\})$, where $p_1 = \ldots = p_4 = 1/4$, such that for any $i, j, k, l \in \{0, 1, 2, 3, 4\}$ with $i + j + k + l = 4$:

$$\mathbb{P}(X_1 = i, X_2 = j, X_3 = k, X_4 = l) = \frac{4!}{i!j!k!l!} p_1^i p_2^j p_3^k p_4^l = \frac{4!}{i!j!k!l!}(1/4)^4$$

The numerical value of the entropy $H(\mathbf{X})$ is resulting directly in bits, as shown in Fig. 4.10.

4.1.6 Dependent Failures

The subject of dependent failures is one of the most relevant issues affecting the validity of standard probabilistic safety analysis methods. This treatment draws on procedures for dealing with common causes as issued by the US Nuclear Regulatory Commission, and the International Agency for Atomic Energy. It is worth to mention that the component data reliability banks typically collect individual component failure events and demands and/or operational times. From such data alone, it is impossible to estimate the probabilities of dependent failures. For this, we need information on the joint failures of components, which becomes available only when incidents involving multiple failures of components are recorded as such. Standard data banks do not collect data on incidents. There is an ongoing programme to analyze the so-called 'Licensee Event Reports (LER)' in the American commercial nuclear power sector, and draw conclusions for probabilistic safety analysis [4].

It is worth to mention here also the International Common Cause Data Exchange (ICDE) project that was initiated by several countries in 1994. The current Phase VII has an agreement period that covers the years 2015–2019. The member countries under the Phase VII Agreement of Organisation for Economic Cooperation and

Development (OECD)/ Nuclear Energy Agency (NEA) and the organizations representing them in the project are as follows: Canada (CNSC), Czech Republic (UJV), Finland (STUK), France (IRSN), Germany (GRS), Japan (NRA), Korea (KAERI), Spain (CSN), Sweden (SSM), Switzerland (ENSI) and the United States (NRC). These countries actually operate 281 NPP units which are about 63% of all NPP units worldwide. With a generation capacity of 275864 MW, these 281 units provide more than 70% of the worlds' total nuclear generation capacity. The number of 281 units comprises 191 PWR, 68 BWR and 23 PHWR so the majority of NPP types is covered.

The ICDE project allows multiple countries to collaborate and exchange Common Cause Failure (CCF) data to enhance the quality of risk analyses, which include CCF modelling. As CCF events are typically rare, most countries do not experience enough CCF events to perform meaningful analyses. Data combined from several countries, however, have yielded sufficient data for more rigorous analyses. The ICDE project has meanwhile published eleven reports on the collection and analysis of CCF events of specific component types (centrifugal pumps, emergency Diesel generators, motor operated valves, safety and relief valves, check valves, circuit breakers, level measurement, control rod drive assemblies and heat exchangers).

A CCF event is defined as a dependent failure in which two or more component fault states exist simultaneously, or within a short time interval, and are a direct result of a shared cause.

Topical reports have been performed or are under preparation [5] for a number of topics, such as external factors, emergency Diesel generators all affected, plant modifications, improving testing, multiunit events and pre-initiator human failure ICDE events.

4.2 Logical Structures

Probabilistic Safety Assessment (PSA) is an established technique to numerically quantify risk measures in chemical, petrochemical or nuclear installations, as well as in certain aerospace applications. It sets out to determine what undesired scenarios can occur, with which likelihood, and what the consequences could be. In addition, it can produce indirect information such as the importance of individual risk contributors.

In the nuclear industry, PSA is required to fulfil the following principal objectives:

1. Provide an estimate of the Core Damage Frequency (CDF) and identify the major accident sequences;
2. Identify those components or plant systems whose unavailability significantly contribute to the core damage frequency;
3. Identify any functional, spatial and human induced dependencies within the plant configuration which contribute significantly to the core damage frequency;
4. Provide a computerized model of the nuclear power plant;

5. Rank the accidence sequences and components according to their relative importance;
6. Evaluate the plant operating experience;
7. Evaluate the plant technical specification and limiting condition of operation;
8. Support decisions on backfitting and design modifications.

PSA comprises a huge model of the nuclear power plant, in which all safety relevant systems, involving thousands of components, are modelled in terms of their reliability and are logically linked together to determine the overall likelihood of core melt accidents or other major accidents.

The logical links are ensured through two main structures: event trees and fault trees. Both methodologies give rise to a pictorial representation of a statement in Boolean logic.

We shall concentrate on fault tree analysis, but briefly explain the difference in the situations modelled by event trees and fault trees. Event trees use 'forward logic' (inductive), whilst the fault trees use 'backward logic' (deductive). An event tree begins with an initiating event (an incident) and 'propagate' this event through the system under study by considering all possible ways in which it can affect the behaviour of the (sub)system.

A such event tree structure is presented in Fig. 4.11.

Terms used to describe the event tree structure are illustrated in the figure and defined below.

- branch—An event associated with the preceding node, usually designated by a point. Mathematically it represents a subset of the sample space for all possible outcomes associated with boolean variables;
- branch probability—The probability of the event represented by the branch conditioned on the occurrence of the events to its left in the event tree;
- end node—The outcome of a pathway belonging to the last level of branches in an event tree. An end node defines a possible end state for a sequence of events;

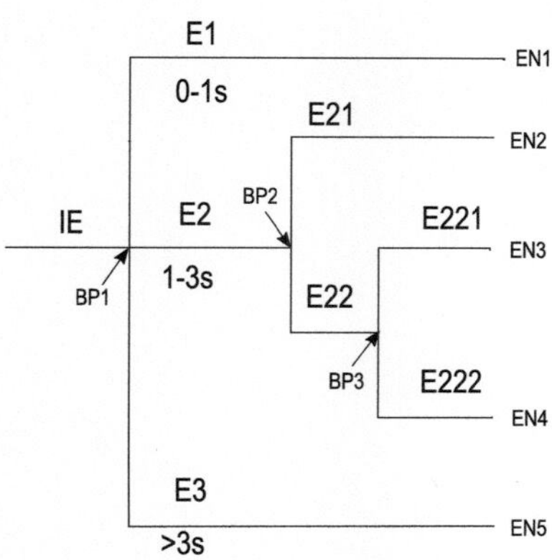

Fig. 4.11 Event tree terminology: IE—initiating event; BP1, BP2, BP3—branch points; E1, E2, E21, E22, E221, E222, E3—events labelled the branches; EN1-EN6—end nodes; $IE \rightarrow E2 \rightarrow E22 \rightarrow E221 \rightarrow EN3$—a pathway

- pathway—A unique sequence of events representing a possible set of events; Mathematically it is the chain of random variable outcomes represented by the intersection of the events along the pathway.

For probabilistic safety assessment of dynamic systems, we cannot usually treat subsystems independently due to the dependencies, such as the time evolution of physical parameters and the changes in the states of (sub)systems. Therefore, we are facing two solutions: one is to look for a way to represent certain branching points of an event tree as logically linked to a shared event of a fault tree/shared state(s) space graph; the other to model the time-dependencies in the form of a series of event trees, where each event tree was related to a specific time instant.

The first solution is illustrated by a hypothetical overspeed turbine-generator failure example shown in Fig. 4.12. The Initiating Event (IE) is a turbine trip signal, following the action of one of the generator protection. The events E1 and E2 are assigned to the separation/isolation of the turbine. The event E3 is a destructive overspeed turbine-generator due to the failure of separation in due time, i.e. the first three seconds following the trip signal. As shown in Fig. 4.12, it is worth to mention that E1 implies an automatic separation/isolation of the turbine, whilst E2 implies a manual separation—based on the action of the main room operator.

The logical structure $M1$ is a Directed Acyclic Graph (DAG) that models the failure of the overspeed protection controller. The logical structure $M2$ is a Direct Graph (DG) that models the states of the stop valves: A—operational state; B, C—non-operational states (one or more governor/interceptor valves stick open or fail to block the flow of steam). The end nodes are as following: EN1—normal state; EN2, EN3, EN4—designed overspeed trip points/states; EN5—destructive overspeed state.

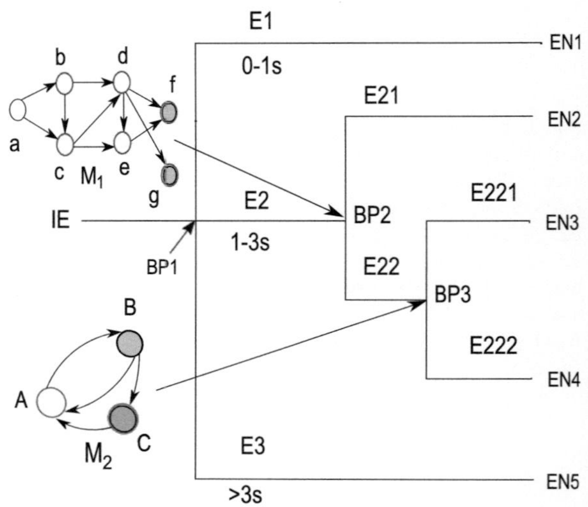

Fig. 4.12 Linking Directed Graphs and Event Tree

The boolean combinations are

$EN1 = IE \star E1$

$EN2 = IE \star E2 \star E21 \star a \star (b \star d \star f \dot{+} c \star d \star f \dot{+} c \star e \star f)$

$EN3 = IE \star E2 \star E22 \star E221 \star a \star (b \star d \star g \dot{+} c \star d \star g)$

$EN4 = IE \star E2 \star E22 \star E222 \star (B \dot{+} C)$

$EN5 = IE \star E3.$

For a more detailed discussion about the subject of linking directed graphs and event trees in PSA studies, see [4, 6].

It worth to mention that in the development and application of Levels 1, 2 and 3 PSA, we need apriori to set—as Level 0—the Initiating Events (IE) and their expected frequencies. We expect that certain End Nodes (EN) of the master Event Trees (ET) will be linked with certain master Fault Trees (FT), as basic events—inputs in fault trees.

The logical structures, such as ET, FT or DAG, are developed and interconnected via several layers, as shown in Fig. 4.13.

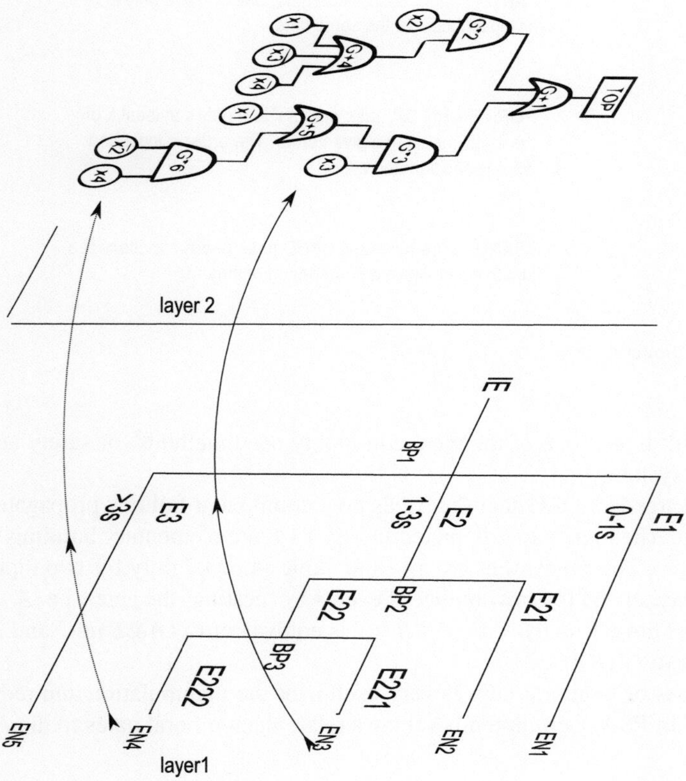

Fig. 4.13 The layers of logical structures

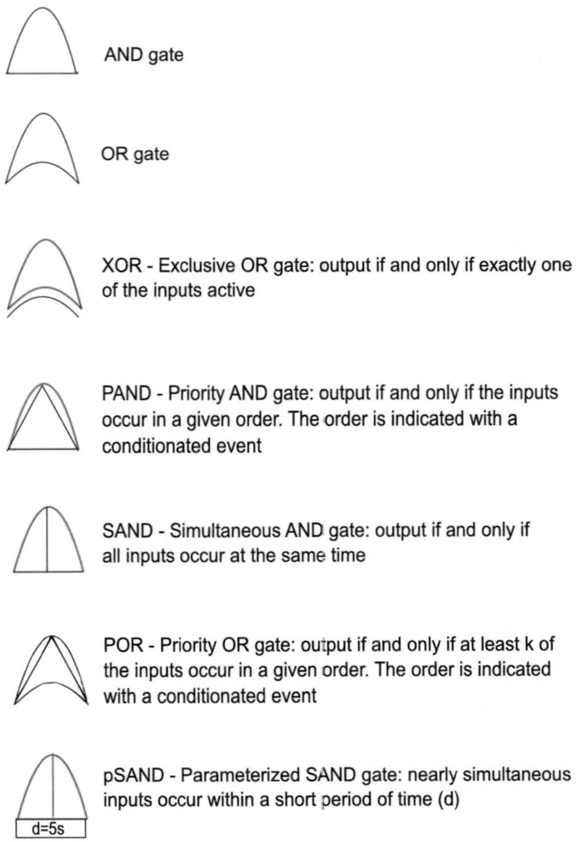

Fig. 4.14 Common gates

The fault tree is one of the most commonly used methods for safety analysis of industrial systems.

A fault tree is a DAG that describes how component failures propagate through the system. The logic gates, depicted in Fig. 4.14, are elementary building blocks of the fault tree. Their meanings are given in Tables 4.6, 4.7 only for two input events, but can be extended for any number of events by 'nesting' the gates, i.e. $A < B < C$ is equivalent to $(A < B) < C$, $A\&B\&C$ is equivalent to $(A\&B)\&C$ and $A|B|C$ is equivalent to $(A|B)|C$.

The class of temporal laws is very useful for the manipulation and reduction of fault trees in PSA. Certain temporal laws relate the temporal gates to the AND, OR gates:

$$X \star Y \iff X < Y \dotplus X\&Y \dotplus Y < X$$

$$X \dotplus Y \iff X|Y \dotplus X\&Y \dotplus Y|X$$

Table 4.6 The symbology for the static gates

Gate	Symbol	Sequence	Meaning
AND	&&	$A\&\&B$	Logical AND ($\star$)
OR	||	$A\|\|B$	Logical OR ($\dot{+}$)
XOR	$\oplus$	$A \oplus B$	Logical XOR

Table 4.7 The symbology for the temporal gates

Gate	Symbol	Sequence	Meaning
PAND	<	$A < B$	A occurs before B occurs. Both A and B must occur
SAND	&	$A\&B$	A occurs at the same time as B occurs. Both A and B must occur
POR	|	$A\|B$	Either A occurs and B does not, or both occur and A occurs first

There are two other laws which can prove useful during reduction: the Laws of POR Transformation and the Laws of Priority [7, 8]. The first are a variant of the Absorption Laws and deal with the Absorption of a POR gate (Table 4.8).

$$X|Y \star Y \iff X < Y$$

$$X|Y \dot{+} Y \iff X \dot{+} Y$$

This behaviour contrasts with the usual behaviour of the temporal gates under Absorption, i.e.:

$$X < Y \star X \iff X < Y$$

$$X < Y \dot{+} X \iff X$$

Table 4.8 The probabilities of the paths shown in Fig. 4.22. The calculus is illustrated in Fig. 4.23

Path	Probability
I	$q_1 \cdot q_2$
II	$p_2 \cdot p_3 \cdot P(4 < 1)$
III	$q_1 \cdot p_2 \cdot q_3 \cdot q_4$
IV	$p_1 \cdot p_2 \cdot P(4\|3)$
V	$p_1 \cdot P(2 < 3)$
VI	$p_1 \cdot P(4 < 3 < 2)$
VII	$p_1 \cdot q_2 \cdot p_3 \cdot q_4$

Fig. 4.15 The probabilities of four types of gates, for two inputs: $X_A \sim Exp[10^{-4}]$, $X_B \sim Exp[2 \cdot 10^{-4}]$

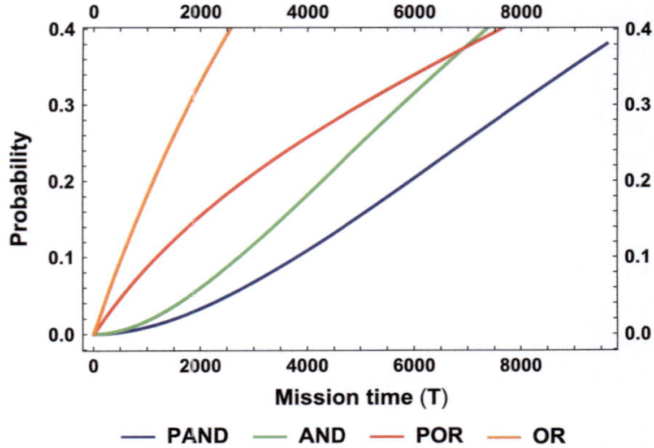

$$X\&Y \star X \iff X\&Y$$

$$X\&Y\dot{+}X \iff X$$

$$X|Y \star X \iff X|Y$$

$$X|Y\dot{+}X \iff X$$

For a feller discussion about the logic and probabilistic quantifications of temporal gates, see [7, 9, 10].

As an example, we show the probabilistic quantifications of temporal gates $PAND$ and POR, respectively, $A < B$ and $A|B$. The graphical representation is presented in Fig. 4.15 and the computer code is described in Fig. 4.16.

Let us denote T the mission time of the nonrepairable system, X_A the occurrence time of event A, X_B the occurrence time of event B and $F_A(x)$ is the cumulative distribution function of the random variable X_A, $f_A(x)$ is the probability density of the random variable X_A and $F_B(x)$ is the cumulative distribution function of the random variable X_B.

Under the hypothesis of statistical independences, thanks to standard inclusion–exclusion formula, the following expressions are true:

- AND

$$P(A\&\&B) = P(0 \leq X_A \leq T \cap 0 \leq X_B \leq T)$$

$$P(A\&\&B) = F_A(T) \cdot F_B(T)$$

- PAND

$$P(A < B) = P(0 < x < X_A \leq x + dx < T \cap x < X_B \leq T | X_B > x)$$

$$P(A < B) = \int_0^T f_A(x) \frac{F_B(T) - F_B(x)}{1 - F_B(x)} dx$$

```
F1 =.;  F2 =.; F3 =.;  F4 =.;  f1 =.;  f2 =.; f3 =.;  f4 =.;
l1 = 0.00001;
l2 = 0.00002;
T =.;
μ3 = μ4 = 11;  σ3 = σ4 = 3;

F1[x_] = CDF[ExponentialDistribution[l1], x];
f1[x_] = PDF[ExponentialDistribution[l1], x];
F2[x_] = CDF[ExponentialDistribution[l2], x];
f2[x_] = PDF[ExponentialDistribution[l2], x];
F3[x_] = CDF[LogNormalDistribution[μ3, σ3], x];
f3[x_] = PDF[LogNormalDistribution[μ3, σ3], x];
F4[x_] = CDF[LogNormalDistribution[μ4, σ4], x];
f4[x_] = PDF[LogNormalDistribution[μ4, σ4], x];

PAND41[T_?NumericQ] :=
  NIntegrate[f4[x] (F1[T] - F1[x]) / (1 - F1[x]), {x, 0, T}];
PAND43[T_?NumericQ] :=
  NIntegrate[f4[x] (F3[T] - F3[x]) / (1 - F3[x]), {x, 0, T}];
PAND23[T_?NumericQ] :=
  NIntegrate[f2[x] (F3[T] - F3[x]) / (1 - F3[x]), {x, 0, T}];
PAND32[T_?NumericQ] :=
  NIntegrate[f3[x] (F2[T] - F2[x]) / (1 - F2[x]), {x, 0, T}];
PAND432[T_?NumericQ] :=
  NIntegrate[PAND43[x] (F2[T] - F2[x]) / (1 - F2[x]), {x, 0, T}];
POR43[T_?NumericQ] := PAND43[T] + NIntegrate[f4[x] (1 - F3[T]), {x, 0, T}];

q1 := F1[T]; q2 := F2[T]; q3 := F3[T]; q4 := F4[T];
p1 = 1 - q1;  p2 = 1 - q2;  p3 = 1 - q3;  p4 = 1 - q4;

PI = q1 q2;

PII = p2 p3 PAND41[T];

PIII = q1 p2 q3 q4 ;

PIV = p1 p2 POR43[T] ;

PV = p1 PAND23[T];

PVI = p1 PAND432[T];

PVII = p1 q2 p3 q4;

PTOP =.;

PTOP[T_] = PI + PII + PIII + PIV + PV + PVI + PVII;
```

Fig. 4.16 Mathematica code illustrated the probabilistic quantifications of the temporal gates PAND and POR

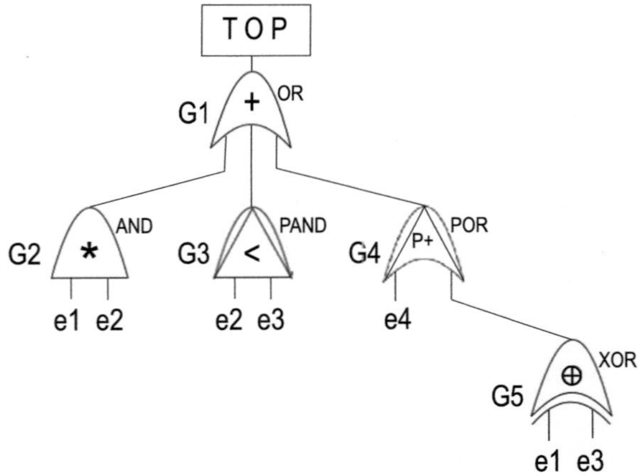

Fig. 4.17 An hypothetical fault tree with dynamic features

- OR

$$P(A||B) = P(0 \leq X_A \leq T \cup 0 \leq X_B \leq T)$$

$$P(A||B) = F_A(T) + F_B(T) - F_A(T) \cdot F_B(T)$$

- POR

$$P(A|B) = P((0 < x < X_A \leq x + dx < T \cap X_B > T) \cup (A < B))$$

$$P(A|B) = F_A(T) \cdot (1 - F_B(T)) + P(A < B)$$

To illustrate the practical significance of the temporal gates, consider the following example that models the reliability of a safety system, with nonrepairable components during the mission time, through an hypothetical fault tree with dynamic features presented in Fig. 4.17.

We first introduce the events $e1$, $e2$, $e3$ and $e4$ as statistically independent random events. For instance, $e1$ and $e2$ might be exponentially distributed and $e3$ and $e4$ might be non-exponentially distributed.

The event TOP is the event that the system is failing to operate in any phase of the mission.

The Sequence Binary Decision Diagrams (SeqBDD) [11] are inspired by the traditional Binary Decision Diagrams (BDD) and applied to analyze fault trees with dynamic features. The main idea is to replace each dynamic gate with its corresponding cut sequence which will be treated as a sequential Boolean variable in the following generating algorithm (Figs. 4.18, 4.19, 4.20, 4.21 and 4.22). The Mathematica code is presented in Fig. 4.23 and the temporal evolution of the top-event

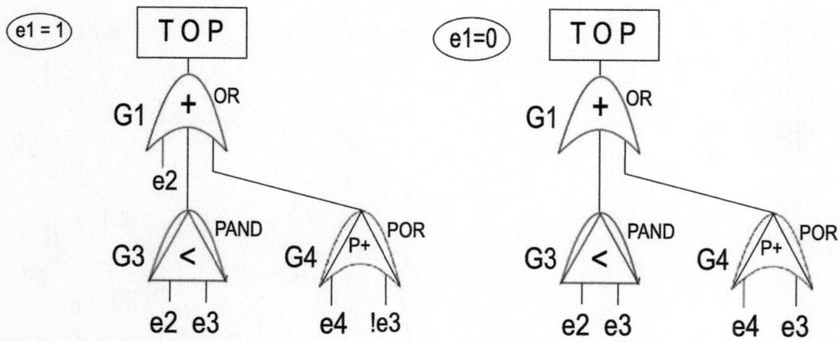

Fig. 4.18 Shannon decomposition of the fault tree with dynamic features: the case e1 $= 1$ (true) on the left side; the case e1 $= 0$ (false) on the right side

Fig. 4.19 Shannon decomposition of the case e1 $= 1$: the case e2 $= 1$ on the left side; the case e2 $= 0$ in the middle (e3 $= 0$) and on the right side (e2 $= 0$; e3 $= 1$)

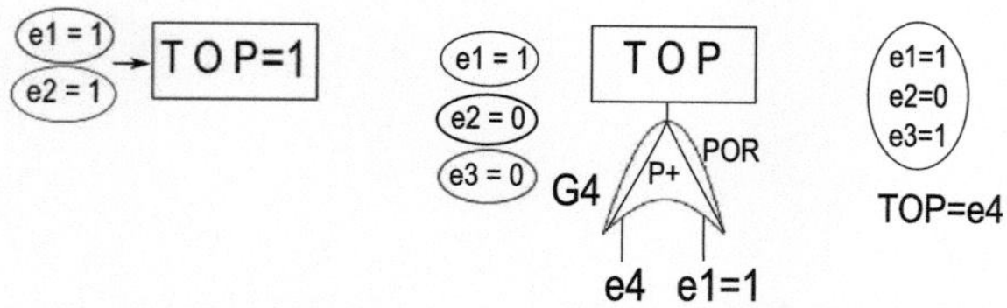

Fig. 4.20 Shannon decomposition of the case e1 $= 0$: the case e2 $= 1$ on the left side; the case e2 $= 0$ on the right side

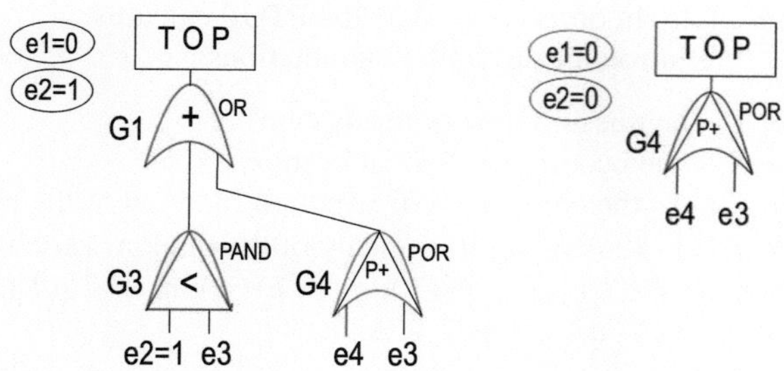

Fig. 4.21 The Sequence Binary Decision Diagram for the hypothetical fault tree with dynamic features

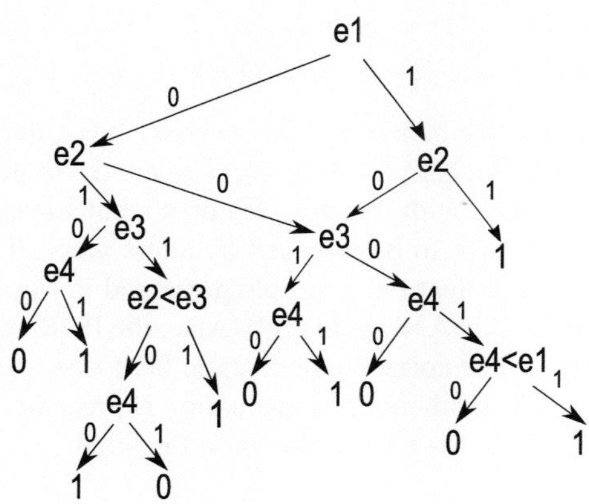

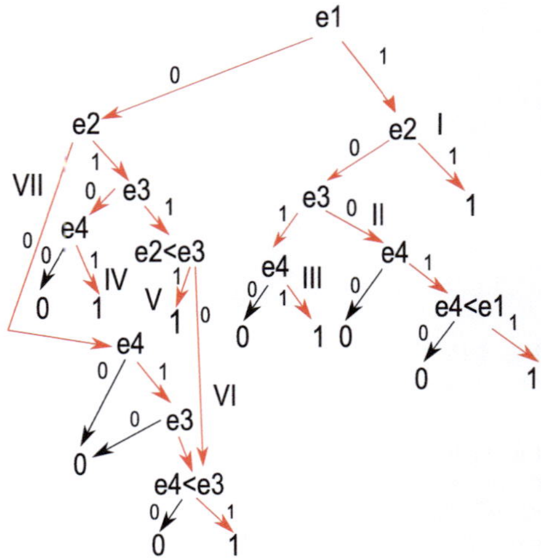

Fig. 4.22 Seven Paths in Sequence Binary Decision Diagram showing the sequences leading to the occurrence of TOP event

probability PTOP is showing in Fig. 4.24. A Monte Carlo simulation is proposed in Fig. 4.25, in order to validate the PTOP calculus.

We introduce the following notations:

- T - the mission time of the system
- X_i - the occurrence time of event e_i
- $F_i(x)$ - the cumulative distribution function of the random variable X_i
- $f_i(x)$ - the probability density of the random variable X_i
- $q_1 = F_1(T); q_2 = F_2(T); q_3 = F_3(T); q_4 = F_4(T)$
- $p_i = 1 - q_i \quad i = 1, 2, 3, 4$
- $P(2 < 3) = \int_0^T f_2(x)(F_3(T) - F_3(x)/(1 - F_3(x))dx$
- $P(3 < 2) = \int_0^T f_3(x)(F_2(T) - F_2(x))/(1 - F_2(x))dx$
- $P(4 < 1) = \int_0^T f_4(x)(F_1(T) - F_1(x))/(1 - F_1(x))dx$

For the benefit of this analysis, the reader interested in looking further is referred to check [12, 13]. We present below the trade-off between the static fault tree and the fault tree with dynamic features results in our example. In this respect, the priority gates have to be replaced by static gates. Thus, in our hypothetical fault tree with dynamic features example presented in Fig. 4.17, the PAND gate has been replaced by a static AND gate (G3). Also, the POR gate has been replaced by a static OR gate (G4). The correspondent static fault tree is proposed in Fig. 4.26.

The mathematical evaluation is presented in Fig. 4.27 and the results are graphically compared and shown in Fig. 4.28.

```
F1 =.; F2 =.; F3 =.; F4 =.; f1 =.; f2 =.; f3 =.; f4 =.;
l1 = 0.00001;
l2 = 0.00002;
T =.;
μ3 = μ4 = 11; σ3 = σ4 = 3;

F1[x_] = CDF[ExponentialDistribution[l1], x];
f1[x_] = PDF[ExponentialDistribution[l1], x];
F2[x_] = CDF[ExponentialDistribution[l2], x];
f2[x_] = PDF[ExponentialDistribution[l2], x];
F3[x_] = CDF[LogNormalDistribution[μ3, σ3], x];
f3[x_] = PDF[LogNormalDistribution[μ3, σ3], x];
F4[x_] = CDF[LogNormalDistribution[μ4, σ4], x];
f4[x_] = PDF[LogNormalDistribution[μ4, σ4], x];

PAND41[T_?NumericQ] :=
  NIntegrate[f4[x] (F1[T] - F1[x]) / (1 - F1[x]), {x, 0, T}];
PAND43[T_?NumericQ] :=
  NIntegrate[f4[x] (F3[T] - F3[x]) / (1 - F3[x]), {x, 0, T}];
PAND23[T_?NumericQ] :=
  NIntegrate[f2[x] (F3[T] - F3[x]) / (1 - F3[x]), {x, 0, T}];
PAND32[T_?NumericQ] :=
  NIntegrate[f3[x] (F2[T] - F2[x]) / (1 - F2[x]), {x, 0, T}];
PAND432[T_?NumericQ] :=
  NIntegrate[PAND43[x] (F2[T] - F2[x]) / (1 - F2[x]), {x, 0, T}];
POR43[T_?NumericQ] := PAND43[T] + NIntegrate[f4[x] (1 - F3[T]), {x, 0, T}];

q1 := F1[T]; q2 := F2[T]; q3 := F3[T]; q4 := F4[T];
p1 = 1 - q1; p2 = 1 - q2; p3 = 1 - q3; p4 = 1 - q4;

PI = q1 q2;

PII = p2 p3 PAND41[T];

PIII = q1 p2 q3 q4 ;

PIV = p1 p2 POR43[T] ;

PV = p1 PAND23[T];

PVI = p1 PAND432[T];

PVII = p1 q2 p3 q4;

PTOP =.;

PTOP[T_] = PI + PII + PIII + PIV + PV + PVI + PVII;
```

Fig. 4.23 Mathematica code illustrated the PTOP calculation based on the seven paths in SeqBDD

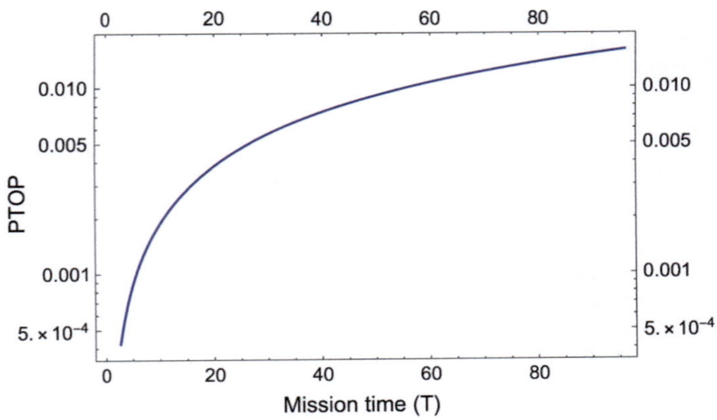

Fig. 4.24 The top event probability PTOP of the fault tree with dynamic features shown in Fig. 4.17

```
T = 120; NS = 1000000; nTOP = 0;
For[i = 1, i <= NS, i++, X1 = RandomVariate[ExponentialDistribution[l1]];

X2 = RandomVariate[ExponentialDistribution[l2]];

X3 = RandomVariate[LogNormalDistribution[μ3, σ3]];

X4 = RandomVariate[LogNormalDistribution[μ4, σ4]];

If[X1 ≤ T && X2 ≤ T, G2 = True, G2 = False];
If[X2 < T && X3 ≤ T && X2 < X3, G3 = True, G3 = False];
 If[X1 < X4 && X1 ≤ T, EventX1 = True, EventX1 = False];
 If[X3 < X4 && X3 ≤ T, EventX3 = True, EventX3 = False];
 EventXOR13 = Xor[EventX1, EventX3];
 If[X4 ≤ T && ! EventXOR13, G4 = True, G4 = False];
TOP = G2 || G3 || G4; If[TOP, nTOP++]]
N[nTOP / NS]
0.019323

PTOP[120]
0.019023
```

Fig. 4.25 Mathematica code illustrated a Monte Carlo simulation validating the PTOP calculus in the case of fault tree with dynamic features

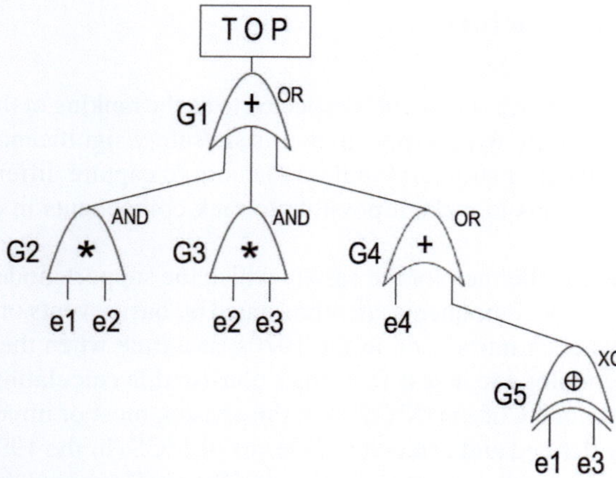

Fig. 4.26 The static fault tree

```
In[170]:= TOPstatic = e1 && e2 || e2 && e3 || e4 || Xor[e1, e3];

expansionRuleN = {! a_ → (1 - a)};
expansionRule2a = {a_ && b_ → a * b};
expansionRule3 = {a_ ⊻ b_ → a + b};

Φ = BooleanConvert[TOPstatic, "ESOP"]

e4 ⊻ (e1 && ! e3 && ! e4) ⊻ (e2 && e3 && ! e4) ⊻ (! e1 && ! e2 && e3 && ! e4)

PTOPstatic[T_] =
  Expand[Fold[ReplaceRepeated, Φ, {expansionRule2a, expansionRuleN, expansionRule3}] /.
     e1 → q1 /. e2 → q2 /. e3 → q3 /. e4 → q4];
```

Fig. 4.27 Mathematica code illustrated the PTOP calculus in the case of static fault tree

Fig. 4.28 Failure probability of the safety system modelled through a static fault tree shown in Fig. 4.17 and respectively through a fault tree with dynamic features as presented in Fig. 4.26

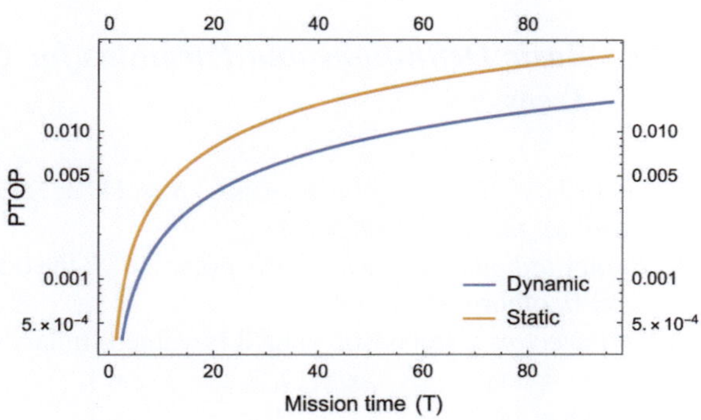

4.3 Importance Factors

One of the activities of risk assessment is expected to be the ranking of the components of the system under study with respect to their risk /safety significance. Importance factors are probabilistic or structural indices that aim to capture different aspects of this significance and thus to make it possible to rank components in different ways [14–16].

They were primarily defined for the case in which the support model is a coherent fault tree and failures of components are represented by basic events of this fault tree. Most of them have been introduced in the 1970s, at a time when the predominant, if not the only, technology to assess fault trees consisted in calculating probabilistic measures from Minimal Cutsets (MCS). For this reason, most of importance factors have been usually defined and calculated in terms of MCS. In the 1990s in the fault tree domain, a new technology came into play: the Binary Decision Diagrams (BDD) [17]. The BDD expresses the failure logic in a Disjoint Normal Form (DNF), which gives it an advantage from the computational viewpoint, especially for large PSA models [18]. An illustration of the Boolean rules and their implementation throughout a BDD structure have been presented in Fig. 4.27. To be fully informative, the line $\Phi = BooleanConvert[TOPstatic, \text{“}ESOP\text{”}]$ is looking for the DNF by calling the argument ESOP (Exclusive Sum of Products). This means a logical sum of disjointed minterms. For instance, in Fig. 4.27, there are nominated four minterms, namely e_4, $e_1 \star \bar{e}_3 \star \bar{e}_4$, $e_2 \star e_3 \star \bar{e}_4$ and $\bar{e}_1 \star \bar{e}_2 \star e_3 \star \bar{e}_4$.

The subject of minterms is a very important one because it has been shown [19] that each importance factor characterizes, in fact, the probability of a certain set of minterms. The notion of critical states, that is, minterms in which failing the component suffices to fail the system, plays a central role in this process.

4.3.1 Basic Definitions and Formulas for Coherent Fault Trees

1. *Set of Fault Tree (FT) basic events*: $S = \{1, 2, \dots, n\}$.
2. *Cardinal number of set A*: $|A|$.
3. *Binary indicator x_i for the basic event $i \in S$*: the occurence of the event i implies $x_i = 0$, otherwise $x_i = 1$.
4. *State vector*: a stochastic valued (s-valued) binary vector $\mathbf{x} = (x_1, x_2, \dots, x_n)$.
5. $(1_i, \mathbf{x}) = (x_1, x_2, \dots, x_{i-1}, 1, x_{i+1}, \dots, x_n)$.
6. $(0_i, \mathbf{x}) = (x_1, x_2, \dots, x_{i-1}, 0, x_{i+1}, \dots, x_n)$.
7. *Basic event $i \in S$ probability*: $q_i = P\{x_i = 0\}$.
8. *Structure function $\Phi(\mathbf{x})$*: is a Boolean function which represents the occurence of the TOP Event (TE) according to the occurences of the basic events of the FT.

9. *Dual structure function* $\Phi^d(\mathbf{x})$: is the logical negation of the structure function $\Phi^d(\mathbf{x})$, i.e. $\Phi^d(\mathbf{x}) = !\Phi(\mathbf{x})$.
10. *TE probability*: $Q = P(\Phi(\mathbf{x}) = 0)$.
11. The basic event i is *irrelevant* to the structure Φ if Φ is constant in x_i, that is $\Phi(1_i, \mathbf{x}) = \Phi(0_i, \mathbf{x})$ for all $\mathbf{x}$. Otherwise, the event i is *relevant* to the structure Φ.
12. A FT is *coherent* if (i) its structure function is nondecreasing and (ii) each basic event is relevant.
13. A *cut* vector is a state vector $\mathbf{x}$ such that $\Phi(\mathbf{x}) = 0$.
14. A *path* vector is a state vector $\mathbf{x}$ such that $\Phi(\mathbf{x}) = 1$.
15. A *minimal cut* vector is a cut vector $\mathbf{x}$ such that $\Phi(\mathbf{y}) = 1$ for all $\mathbf{y} \geq \mathbf{x}, \mathbf{y} \neq \mathbf{x}$.
16. A *minimal path* vector is a path vector $\mathbf{x}$ such that $\Phi(\mathbf{y}) = 0$ for all $\mathbf{y} \leq \mathbf{x}, \mathbf{y} \neq \mathbf{x}$.
17. set of cuts: $\mathscr{C} = \{C_1, C_2, \dots C_u\}$.
18. set of cuts of size d: $\mathscr{C}(d)$.
19. set of cuts containing the basic event $i \in S$: $\mathscr{C}_i$.
20. set of cuts of size d containing the basic event $i \in S$: $\mathscr{C}_i(d)$.
21. set of cuts not containing the basic event $i \in S$: $\mathscr{C}_{(i)}$.
22. set of cuts of size d not containing the basic event $i \in S$: $\mathscr{C}_{(i)}(d)$.
23. set of paths: $\mathscr{P} = \{P_1, P_2, \dots P_v\}$.
24. set of paths of size d: $\mathscr{P}(d)$.
25. set of paths with reference to the basic event $i \in S$: $\mathscr{P}_i$.
26. set of paths with no reference to the basic event $i \in S$: $\mathscr{P}_{(i)}$.
27. $|\mathscr{C}_i| + |\mathscr{P}_{(i)}| = 2^{n-1}$.
28. $|\mathscr{P}_i| + |\mathscr{C}_{(i)}| = 2^{n-1}$.
29. $|\mathscr{C}_i(d)| + |\mathscr{P}_{(i)}(n-d)| = C_{n-1}^{d-1}$.
30. $|\mathscr{P}_i(d)| + |\mathscr{C}_{(i)}(n-d)| = C_{n-1}^{d-1}$.
31. $|\mathscr{P}_i| - |\mathscr{P}_{(i)}| = |\mathscr{C}_i| - |\mathscr{C}_{(i)}|$.

Example. Let us consider a fault tree with $n = 3$ basic events, namely 1, 2 and 3. We define the s-valued binary vector $\mathbf{x} = (x_1, x_2, x_3)$. The occurence of the event i implies $x_i = 0$, otherwise $x_i = 1, i = 1, 2, 3$. The structure function of this fault tree is $\Phi(\mathbf{x}) = !x_1 \star !x_2 \dot{+} !x_1 \star !x_3 \dot{+} !x_2 \star !x_3$. The dual structure function of this fault tree is $\Phi^d(\mathbf{x}) = !\Phi(\mathbf{x}) = (x_1 \dot{+} x_2) \star (x_1 \dot{+} x_3) \star (x_2 \dot{+} x_3) = x_1 \star x_2 \dot{+} x_1 \star x_3 \dot{+} x_2 \star x_3$. Consequently, the set of cuts is $C_1 = \{!1, !2\}$, $C_2 = \{!1, !3\}$, $C_3 = \{!2, !3\}$, $C_4 = \{!1, !2, !3\}$ and the set of paths is $P_1 = \{1, 2\}$, $P_2 = \{1, 3\}$, $P_3 = \{2, 3\}$, $P_4 = \{1, 2, 3\}$. With reference to basic event 1, we have $\mathscr{C}_1 = \{C_1, C_2, C_4\}$, respectively, $\mathscr{P}_{(1)} = \{P_3\}$. Therefore, $|\mathscr{C}_1| = 3$, $|\mathscr{P}_{(1)}| = 1$ and $|\mathscr{C}_1| + |\mathscr{P}_{(1)}| = 2^{3-1}$.

32. Basic event/component i is critical for coherent structure $\Phi(\mathbf{x})$ at the state vector $(\cdot_i, \mathbf{x})$ when $\Phi(1_i, \mathbf{x}) = 1$ and $\Phi(0_i, \mathbf{x}) = 0$.
33. $(\cdot_i, \mathbf{x})$ is a critical vector for basic event/component i, if and only if there exists a minimal path P_r and a minimal cut C_k such that (i) $P_r \cap C_k = \{i\}$ and (ii) $x_j = 1$ for all $j \in P_r \backslash \{i\}$, and $x_j = 0$ for all $j \in C_k \backslash \{i\}$.

34. The set of state vectors $(\cdot_i, \mathbf{x})$ in which basic event / component i is critical for the system: $CR(i) = \{(\cdot_i, \mathbf{x}) : \Phi(1_i, \mathbf{x}) - \Phi(0_i, \mathbf{x}) = 0\}$.
35. The number of critical vector for basic event/component i: $n_\varphi(i) = |CR(i)| = \sum_{\mathbf{x}}(\Phi(1_i, \mathbf{x}) - \Phi(0_i, \mathbf{x}))$.
36. For a coherent fault tree (S, Φ) with basic event probability vector $\mathbf{q} = (q_1, q_2, \ldots, q_n)$ and dual structure function $\Phi^d =!\Phi$ with component reliability/availability vector $\mathbf{p} = (p_1, p_2, \ldots, p_n)$, where $q_i = 1 - p_i$, for $i = 1, 2, \ldots, n$, let $I_B(i, \Phi, \mathbf{q})$, $I_B(i, \Phi^d, \mathbf{p})$ be the Birnbaum probabilistic importance factors of basic event/component i. Let us also denote $P = \mathbb{P}\{\Phi^d(\mathbf{x}) = 1\}$, $Q = \mathbb{P}\{\Phi(\mathbf{x}) = 0\}$, where $P + Q = 1$.
    Consider

$$I_B(i, \Phi, \mathbf{q}) = \mathbb{P}\{\Phi(1_i, \mathbf{x}) = 0\} - \mathbb{P}\{\Phi(0_i, \mathbf{x}) = 0\} = \partial Q/\partial q_i$$

$$I_B(i, \Phi^d, \mathbf{p}) = \mathbb{P}\{\Phi^d(1_i, \mathbf{x}) = 1\} - \mathbb{P}\{\Phi^d(0_i, \mathbf{x}) = 1\} = \partial P/\partial p_i$$

    Thus, for a coherent system, we have

$$I_B(i, \Phi, \mathbf{q}) = I_B(i, \Phi^d, \mathbf{p})$$

37. The Birnbaum probabilistic factor is given by the formula

$$I_B(i) = I_B(i, \Phi, \mathbf{q}) = I_B(i, \Phi^d, \mathbf{p})$$

38. Let us denote $PCR(i) = \mathbb{P}\{(\cdot_i, \mathbf{x}) \in CR(i)\}$.
    Then

$$p_i \cdot I_B(i) = PCR(i)$$

39. A formation of S is a set of minimal paths $\mathscr{P}$ whose union is S.
40. The signed domination $d(\mathscr{P})$ of (S, Φ^d) is the number of odd formations of S minus the number of even formation of S.
41. The domination $D(\mathscr{P})$ of (S, Φ^d) is $|d(\mathscr{P})|$.
42. For a given component $i \in S$, $\mathscr{P}_{-i}$ are the min paths of $\mathscr{P}$ not containing i while $\mathscr{P}_{+i}$ is obtained by deleting i from all min paths of $\mathscr{P}$ and then discarding any superset which may now be present.
43. The signed domination theorem: for all coherent systems, with independent components

$$d(\mathscr{P}) = d(\mathscr{P}_{+i}) - d(\mathscr{P}_{-i})$$

    Example. The following example of an undirected network will illustrate ideas. For this example $S = \{1, 2, 3, 4, 5\}$ while $\mathscr{P} = \{\{1, 2\}, \{3, 4\}, \{2, 3, 5\}, \{1, 4, 5\}\}$. The reliability polynomial for this example is $2r^2 + 2r^3 - 5r^4 + 2r^5$. The formations of S are

$F_0 = \{\{1, 2\}, \{3, 4\}, \{2, 3, 5\}, \{1, 4, 5\}\}$.
$F_1 = \{\{2, 3, 5\}, \{1, 4, 5\}\}$.
$F_2 = \{\{1, 2\}, \{3, 4\}, \{2, 3, 5\}\}$.
$F_3 = \{\{1, 2\}, \{3, 4\}, \{1, 4, 5\}\}$.
$F_4 = \{\{3, 4\}, \{2, 3, 5\}, \{1, 4, 5\}\}$.
By inclusion–exclusion formula, the coefficient of r^n is the number of odd formations minus the number of even formations, i.e. the signed domination value $d(\mathscr{P})$.
In our example, the number of odd formations is three, namely F_2, F_3 and F_4. The number of even formations is two, namely F_0 and F_1, so the coefficient of r^n is 2.
Another example. Let us consider a fault tree with $n = 5$ basic events, namely 1, 2, 3, 4 and 5. We define the s-valued binary vector $\mathbf{x} = (x_1, x_2, x_3, x_4, x_5)$. The occurence of the event i implies $x_i = 0$, otherwise $x_i = 1, i = 1, \ldots, 5$. The set of minimal paths is $P_1 = \{1, 2\}$, $P_2 = \{3, 4\}$, $P_3 = \{2, 3, 5\}$, $P_4 = \{1, 4, 5\}$. The dual structure function is $\Phi^d(\mathbf{x}) = x_1 \star x_2 \dot{+} x_3 \star x_4 \dot{+} x_2 \star x_3 \star x_5 \dot{+} x_1 \star x_4 \star x_5$. The structure function of this fault tree is resulting $\Phi(\mathbf{x}) = !\Phi^d(\mathbf{x}) = !x1 \star !x3 \dot{+} !x2 \star !x4 \dot{+} !x1 \star !x4 \star !x5 \dot{+} !x2 \star !x3 \star !x5$.
The set of minimal cuts is $C_1 = \{!1, !3\}$, $C_2 = \{!2, !4\}$, $C_3 = \{!1, !4, !5\}$, $C_4 = \{!2, !3, !5\}$. With regard to basic event / component 1, the minimal paths and minimal cuts that contain the literal 1 are P_1, P_4, C_1 and C_3. Therefore, the critical vectors for 1, namely $(\cdot_1, \mathbf{x})$, are resulting following again the above-mentioned theorem (33):

- $P_1 \cap C_1 = 1 \rightarrow (1, 1, 0, *, *) : (1, 1, 0, 1, 1); (1, 1, 0, 1, 0); (1, 1, 0, 0, 1); (1, 1, 0, 0, 0)$.
- $P_1 \cap C_3 = 1 \rightarrow (1, 1, *, 0, 0) : (1, 1, 1, 0, 0); (1, 1, 0, 0, 0)$.
- $P_4 \cap C_1 = 1 \rightarrow (1, *, 0, 1, 1) : (1, 1, 0, 1, 1); (1, 0, 0, 1, 1)$.

We find $n_\varphi(1) = 6$, i.e.

$$\begin{aligned} CR(1) = &\{(1, 1, 0, 1, 1); (1, 1, 0, 1, 0); (1, 1, 0, 0, 1)\} \cup \\ &\{(1, 1, 0, 0, 0); (1, 1, 1, 0, 0); (1, 0, 0, 1, 1)\}. \end{aligned}$$

With regard to basic event/component 5, the minimal paths and minimal cuts that contain the literal 5 are P_3, P_4, C_3 and C_4. Therefore, the critical vectors for 5, namely $(\cdot_5, \mathbf{x})$, are resulting following the above-mentioned theorem (33):

- $P_3 \cap C_3 = 5 \rightarrow (0, 1, 1, 0, 1)$.
- $P_4 \cap C_4 = 5 \rightarrow (1, 0, 0, 1, 1)$.

We find $n_\varphi(5) = 2$, i.e. $CR(5) = \{(0, 1, 1, 0, 1); (1, 0, 0, 1, 1)\}$.
Thus, $PCR(5) = q_1 \cdot p_2 \cdot p_3 \cdot q_4 \cdot p_5 + p_1 \cdot q_2 \cdot q_3 \cdot p_4 \cdot p_5$
and $I_B(5) = PCR(5)/p_5 = q_1 \cdot p_2 \cdot p_3 \cdot q_4 + p_1 \cdot q_2 \cdot q_3 \cdot p_4$.

44. Birnbaum's structural importance factor [20]:
Given a basic event/component $i \in S$, we have $I_B^\varphi(i) = n_\varphi(i)/2^{n-1}$.

45. Critical importance factor (Lambert) [19]:

$$CIF(i) = \partial Q/\partial q_i \cdot q_i/Q = q_i \cdot IB(i)/Q$$

It normalizes IB(i) through the ratio of the probability of basic event i and the nominal value of the risk metric Q. CIF enables to discriminate among components that have the same IB(i). Thus, a component less reliable appears to be more critical than a component more reliable, even if both components have the same IB.

46. Risk reduction worth [19]:

$$RRW(i) = Q(1_i, \mathbf{x})/Q = (Q|q_i = 0)/Q$$

It measures the amount that the TE probability would decrease assuming that the basic event i never occurs.

47. Risk achievement worth [19]:

$$RAW(i) = Q(0_i, \mathbf{x})/Q = (Q|q_i = 1)/Q$$

It measures the amount that the TE probability would increase if the basic event i happens almost surely.

48. Fussell–Vesely [19]:

$$FV(i) = P(\mathcal{C}_i)/Q = P(x_i = 0|TOP\ Event\ occurred)$$

It measures the overall percent contribution of cut sets containing the basic event i of interest to the total Top Event (TE) probability, i.e. Q.

49. Barlow–Proschan [19]:

$$I_{BP}(i) = \int_0^1 (Q_{q_i=1} - Q_{q_i=0})|_{q_1=q_2=\ldots q_{i-1}=q_{i+1}=\ldots q_n=q} dq$$

50. Differential importance measure [19]:
Differential importance measure (DIM) considers the total variation of the TE probability Q due to a small variation of its parameters, taken one at a time. If the variation of the parameter is small enough, the variation of Q is the total differential dQ:

$$dQ = \sum_{k=1}^{n} (\partial Q/\partial q_k) dq_k$$

The DIM of the basic event i, $DIM(i)$, is defined as the fraction of the total change in Q which pertains to the change in the parameter q_i:

$$DIM(i) = \frac{dQ_i}{dQ} = \frac{(\partial Q/\partial q_i) dq_i}{\sum_{k=1}^{n} (\partial Q/\partial q_k) dq_k} = \frac{IB(i) dq_i}{\sum_{k=1}^{n} IB(k) dq_k}$$

51. DIM is additive in the sense that the DIM of a subset of basic events, let's say s, $t, \ldots, w$, is

$$DIM(s,t,\ldots,w) = \frac{dQ_{s,t,\ldots,w}}{dQ} = DIM(s) + DIM(t) + \cdots + DIM(w)$$

52. The relationships between RAW, RRW, BI and CIF are based on their definitions for PSA models:

$$RAW(i) - RRW(i) = ((Q|q_i = 1) - (Q|q_i = 0))/Q = IB(i)/Q$$

$$IB(i) = Q \cdot (RAW(i) - RRW(i))$$

$$CIF(i) = IB(i) \cdot q_i/Q = q_i \cdot (RAW(i) - RRW(i))$$

$$RAW(i) = RRW(i) + IB(i)/Q$$

$$RRW(i) = RAW(i) - CIF(i)/q_i$$

Extension of some importance factors for basic event groups.
In many cases, it is of interest to evaluate the importance of a set of basic events instead of just individual basic events.

53. Joint failure importance [21]:
For the basic events a and b, the joint failure importance (JFI) is defined as

$$JFI(a,b) = \frac{d^2Q}{dq_a dq_b}$$

For the basic events $i_1, i_2, \ldots i_k$, the joint failure importance (JFI) is defined as

$$JFI(i_1, i_2, \ldots, i_k) = \frac{d^kQ}{dq_{i_1} dq_{i_2} \ldots dq_{i_k}}$$

54. The FV of a subset of basic events, i.e. a basic event group, can be found by extending its definition as

$$FV(i_1, i_2, \ldots, i_k) = P(\mathscr{C}_{i_1} \cup \mathscr{C}_{i_2} \cup \cdots \cup \mathscr{C}_{i_k})/Q$$

It measures the overall percent contribution of cut sets containing the basic events $i_1, i_2, \ldots i_k$ of interest to the total TE probability, i.e. Q.

References

1. IEC (2010) IEC 61508-6:2010 - Functional safety of electrical/electronic/programmable electronic safety- related systems - Part 6: Guidelines on the application of IEC 61508-2 and IEC 61508-3
2. ISA (2015) ISA- TR84.00.02-2015 - Safety Integrity Level (SIL) Verification of Safety Instrumented Functions
3. SINTEF (2013) Reliability Prediction Method for Safety Instrumented Systems - PDS Method Handbook. A24442, Trondheim
4. Xu H, Dugan JB (2004) Combining dynamic fault trees and event trees for probabilistic risk assessment. In: Annual Symposium Reliability and Maintainability, 2004 - RAMS, pp 214–219. https://doi.org/10.1109/RAMS.2004.1285450
5. Agency NE (2018) NEA / CSNI report. www.oecd-nea.org/nsd/docs/indexcsni.html
6. Andrews JD, Dunnett SJ (2000) Event-tree analysis using binary decision diagrams. IEEE Trans Reliab 49(2):230–238. https://doi.org/10.1109/24.877343
7. Fussell JB, Aber EF, Rahl RG (1976) On the quantitative analysis of priority-and failure logic. IEEE Trans Reliab R-25(5):324–326. https://doi.org/10.1109/TR.1976.5220025
8. Tang Z, Dugan JB (2004) Minimal cut set/sequence generation for dynamic fault trees. In: Annual symposium reliability and maintainability, 2004 - RAMS, pp 207–213. https://doi.org/10.1109/RAMS.2004.1285449
9. (2013) Zamojski W, Mazurkiewicz J, Sugier J, Walkowiak T, Kacprzyk J (eds) (2013) Quantification of simultaneous-AND gates in temporal fault trees. Advances in intelligent systems and computing, vol 224. Springer, New York. https://doi.org/10.1007/978-3-319-00945-2
10. (2012) .1007/978-3-642-33678 Quantification of priority-OR gates in temporal fault trees. Lecture notes in computer science, vol 7612. Springer, New York. https://doi.org/10.1007/978-3-642-33678-2
11. Xing L, Shrestha A, Dai Y (2011) Exact combinatorial reliability analysis of dynamic systems with sequence-dependent failures. Reliab Eng Syst Saf 96(10):1375–1385. https://doi.org/10.1016/j.ress.2011.05.007, http://www.sciencedirect.com/science/article/pii/S0951832011001050
12. Krčál J, Krčál P (2015) Scalable analysis of fault trees with dynamic features. In: 2015 45th Annual IEEE/IFIP international conference on dependable systems and networks, pp 89–100. https://doi.org/10.1109/DSN.2015.29
13. Rauzy A (2015) Towards a sound semantics for dynamic fault trees. Reliab Eng Syst Saf 142:184–191. https://doi.org/10.1016/j.ress.2015.04.017
14. Fussell JB (1975) How to hand-calculate system reliability and safety characteristics. IEEE Trans Reliab R-24(3):169–174. https://doi.org/10.1109/TR.1975.5215142
15. Kuo W, Zhu X (2012) Importance measures in reliability, risk and optimization - principles and applications. Wiley, Chichester
16. Lambert H (1975) Measures of importance of events and cut sets in fault trees. In: Barlow RE, Fussel JB, Singpurwalla ND (eds) Reliability and fault tree analysis. SIAM Press, Philadelphia, pp 77–100
17. Bryant RE (1986) Graph-based algorithms for boolean function manipulation. IEEE Trans Comput C-35(8):677–691. https://doi.org/10.1109/TC.1986.1676819
18. Ulmeanu AP (2012) Analytical method to determine uncertainty propagation in fault trees by means of binary decision diagrams. IEEE Trans Reliab 61(1):84–94. https://doi.org/10.1109/TR.2012.2182812
19. Dutuit Y, Rauzy A (2014) Importance factors of coherent systems: a review. Proc Inst Mech Eng Part O: J Risk Reliab 228(3):313–323. https://doi.org/10.1177/1748006X13512296
20. Birnbaum ZW (1969) On the importance of different components in a multicomponent system. pp 581–592
21. Armstrong MJ (1995) Joint reliability-importance of components. IEEE Trans Reliab 44(3):408–412. https://doi.org/10.1109/24.406574

Index

MIX
Papier aus verantwortungsvollen Quellen
Paper from responsible sources
FSC® C105338

If you have any concerns about our products,
you can contact us on
ProductSafety@springernature.com

In case Publisher is established outside the EU,
the EU authorized representative is:
Springer Nature Customer Service Center GmbH
Europaplatz 3, 69115 Heidelberg, Germany

Printed by Libri Plureos GmbH
in Hamburg, Germany